ISBN 978-1-330-87269-7
PIBN 10055545

1 MONTH OF
FREE
READING

at
www.ForgottenBooks.com

By purchasing this book you are
eligible for one month membership to
ForgottenBooks.com, giving you
unlimited access to our entire
collection of over 1,000,000 titles via
our web site and mobile apps.

To claim your free month visit:

www.forgottenbooks.com/free55545

English
Français
Deutsche
Italiano
Español
Português

www.forgottenbooks.com

Mythology Photography **Fiction**
Fishing Christianity **Art** Cooking
Essays Buddhism Freemasonry
Medicine **Biology** Music **Ancient
Egypt** Evolution Carpentry Physics
Dance Geology **Mathematics** Fitness
Shakespeare **Folklore** Yoga Marketing
Confidence Immortality Biographies
Poetry **Psychology** Witchcraft
Electronics Chemistry History **Law**
Accounting **Philosophy** Anthropology
Alchemy Drama Quantum Mechanics
Atheism Sexual Health **Ancient History**
Entrepreneurship Languages Sport
Paleontology Needlework Islam
Metaphysics Investment Archaeology
Parenting Statistics Criminology
Motivational

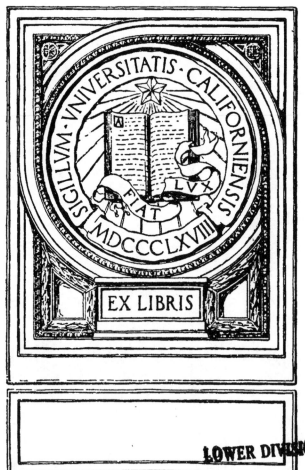

GENERAL PHYSICS

WOOLLCOMBE

London
HENRY FROWDE
Oxford University Press Warehouse
Amen Corner, E.C.

New York
MACMILLAN & CO., 66 FIFTH AVENUE

Practical Work in General Physics

FOR USE IN SCHOOLS AND COLLEGES

Slate

BY

W. G. WOOLLCOMBE, M.A., B.Sc.

SENIOR SCIENCE MASTER IN KING EDWARD'S HIGH SCHOOL, BIRMINGHAM
AUTHOR OF 'PRACTICAL WORK IN HEAT'

Oxford

AT THE CLARENDON PRESS

1894

'Now the forces and action of bodies are circumscribed and measured
either by distances of space, or by moments of time, or by concentration
of quantity, or by predominance of virtue; and, unless these four things
have been well and carefully weighed, we shall have sciences, fair perhaps
in theory, but in practice inefficient.'   *Nov. Org.* ii. Aph. xliv.

INTRODUCTION

—••—

IT is perhaps too late in the day to insist upon the comparatively small value of mere book knowledge in any of the experimental sciences. To reap the full benefit from the mental discipline a scientific study offers reading must go hand in hand with individual experiment. In the case of Chemistry and Biology this is universally acknowledged, but in Physics the experiments are too generally performed by the lecturer himself, which deprives the teaching of much of its value. The cost of providing for experimental work, even in the case of large classes, need not be great. In the present volume as well as in the author's *Practical Work in Heat* an essential feature is to offer a fairly complete experimental course in the ground covered at but a trifling cost. The question of time is always a crux with the authorities who have to arrange the time table, but if a course of Practical Physics were estimated at its proper value this difficulty would be at once met.

The knowledge of physical facts is not the be-all and end-all of the study. The student must be brought up to do the experiments himself, and, unless he has learned by his own experience and observation what experimental evidence means, he will be unable to appreciate rightly the evidence on which is based the reasoning by which the general conclusions of Physical

Science are established. It would not, perhaps, be too much to say that if a student has conscientiously been through even such a course as is here offered to him, he will have derived far more benefit mentally than if he had stored his mind with a mass of facts and formulae out of books.

The advantages of a course of practical physics as a mental discipline are numerous. It teaches us the experimental method and the process of inductive reasoning which it involves, and the result of the training ought to give the student confidence in what he has seen with his own eyes and reasoned out with his own mind.

It also teaches us accurate observation. We all know the story of *Eyes and No Eyes.* Every measurement necessitates primarily a choice of a unit of the same nature as the quantity to be observed, and also the reading of a scale of such units (cf. Thermometer, Barometer, Galvanometer, &c.), and it is only by practice that a student becomes able to read the simplest scale with sufficient accuracy, as any one can testify who has had to do with beginners.

Again, the interest of the student is aroused on the subject if he has the opportunity of handling the apparatus and performing experiments in illustration of his book-work. 'Without observation and experiment,' says Desaguliers, ' our Natural Philosophy would be a science of terms and an unintelligible jargon.' It is only by individual experiments that we can infuse life into what some students are otherwise apt to look upon as the dry bones of science.

From a purely utilitarian point of view a course of practical physics is of great importance. The mere fact of reading a particular scale is in itself of no value, but the consequent sharpening of the faculties of observation

is of prime importance, as the cuteness of a successful
man of business depends much on his observational
powers, and success in Chemistry, Astronomy, Civil and
Electrical Engineering, &c. on a thorough understanding
of the methods of accurate measurement.

Another advantage in a course of this kind is to
teach the student to look for errors he may have
committed and to adopt means to eliminate them.
However precise our observations or instruments, we
never get the same number if we repeat an experiment.
The errors involved may be (*a*) **systematic errors**, over
which we have no control. For instance, the temperature of the room may alter, or the currents of air
may affect the balance differently and may tend to
give different results ; (*β*) **constant errors**, due to the
graduation of the scale not being correct or the zero
of the scale having altered : these, being due to
the instrument used, can only be eliminated by comparing with a standard scale or by observing the zero
error (Note 4) ; (*γ*) **accidental errors**, which may be got
rid of in various ways according to the experiment we
have in hand. The following means of neutralizing
accidental errors may be noted.

(i) *By the method of least error*, i.e. by arranging the
experiment so as to reduce possible errors to a mini-
mum ; e. g. in determining the densities of solids we
take as large a piece of the solid as possible. It is
evident that an error of 1 mgr. in estimating a mass of
100 grs. is only $\frac{1}{100}$th of the same error in estimating the
mass of 1 gr.

(ii) *By the method of multiplication.* For instance,
in finding the time of vibration of a pendulum (Experiment 44), by observing the time of twenty vibrations we
only commit $\frac{1}{20}$th of the error we should have made if
we observed the time of one vibration only.

(iii) *By check methods*, or by repeating an experiment after altering the conditions under which it is performed. Refer to Experiments 5, 31, 36, 44, 47, in which check methods are applied.

(iv) *By the method of averages.* If we repeat an experiment two or three times, assuming all our results to be equally trustworthy, the error probably is in excess in some, in defect in others, so that on taking the average of our results we get a value nearer the true one than either of the individual results.

A student may think that if an apparatus of greater precision were at his disposal he would be relieved of the continual tax which he feels upon his powers of observation. This is, however, not the case. The better the instrument the harder it is to do justice to it. One must learn to get the best possible results with rough instruments before one is fitted to use instruments of precision.

The student should also distinguish between real and apparent accuracy, and should not be deceived by an imposing array of decimal places, nor try to get a nearer accuracy than the instrument he uses will permit (Note 10). Thus, in Experiment 44, if an ordinary watch with a seconds hand is used, we cannot reckon the time of twenty vibrations to a greater accuracy than half a second, which gives a possible error of $\frac{1}{40}$ th of a second for the time of one vibration. Now, in a pendulum of a metre in length, an increase of one centimetre in its length only involves a difference of $\frac{1}{100}$ th of a second in its time of vibration, so that in measuring its length we need only measure to the nearest centimetre. If we lessened the error in measuring the time by using a stop-watch or by taking the time of a greater number of vibrations, we should have to measure the length with greater accuracy.

A word as to the conduct of the classes. Those

under the author's charge average about twenty, and for the juniors one hour a week is devoted to Practical Physics. The students work in pairs, and, as the lectures are arranged so as to be in advance of the practical work, each set can, with the instructions before them, begin work at once, allowing ample time for the teacher to go round and give any further explanation or help in manipulation required. Each student has a note-book devoted specially to his practical work, and, on the completion of an experiment, if the result is satisfactory, he enters into his note-book—

(i) The enunciation of the experiment.

(ii) A neatly-drawn figure of the apparatus used.

(iii) A description in his own words of the method pursued.

(iv) Each measure made, and the results in a properly tabulated form, mere arithmetical calculations only being omitted.

' This last is essential, as, unless insisted upon, a great deal of the good effect is lost.

The author begs to acknowledge his indebtedness chiefly to Harold Whiting's *Physical Measurements*, and ventures to express a hope that this book may be the means of suggesting the adoption of a most important branch of study in all schools, large and small, where science is taught.

King Edward's High School,
Birmingham,
March, 1894.

CONTENTS

—••—

LIST OF EXPERIMENTS.

Those experiments which are marked with an asterisk (*) are suitable only for older students.

E. Archimedes' Principle.

F. Measurement of Volumes.

G. Density or Specific Mass.

PRACTICAL WORK IN GENERAL PHYSICS

————◆◆————

A. Units of Measurement.

In the metric system of units, the unit of length [called
· a *metre*[1]] is the distance between two cross marks on a certain
rod of platinum kept in the French Archives, measured when
the rod is at the temperature of melting ice.   In making this
unit, it was intended that its length should be equal to the
ten-millionth part of the Earth's meridional quadrant passing
through Paris.   The metre is divided into ten equal divisions,
each called a *decimetre* [dcm.] : each decimetre is also divided
into ten equal divisions, each called a *centimetre* [cm.] : each
centimetre is again divided into ten equal divisions, each
called a *millimetre* [mm.].   In physical measurements the metre
is too large a unit, so that all lengths are reckoned in centimetres, and the scales in general use have only centimetres and
millimetres marked upon them.

Table of Lengths.

10 millimetres = 1 centimetre.
10 centimetres = 1 decimetre.
10 decimetres = 1 metre.
1000 metres   = 1 kilometre.

The measurement of areas and volumes is based upon the

---

[1] The metre is very nearly equal to 39.37 inches.

measurement of lengths, as will be seen in the following tables:—

Table of areas.

100 sq. millimetres = 1 sq. centimetre [sq. cm.].
100 sq. centimetres = 1 sq. decimetre.
100 sq. decimetres = 1 sq. metre.

Table of volumes.

1000 cub. millimetres = 1 cub. centimetre [c.c.].
1000 cub. centimetres = 1 cub. decimetre [called a *litre*].
1000 cub. decimetres = 1 cub. metre.

The unit of mass, called a *kilogram*[2], is the mass of a certain piece of platinum kept in the French Archives. In making this unit, it was intended that its mass should be that of a cubic decimetre or litre of distilled water at 4° C. Kupffer has, by a series of accurate experiments, proved that the true mass of a cubic decimetre of distilled water at 4° C is 1·000013 kilogram, so that for practical purposes we may take the kilogram to be of the value originally intended. In physical determinations the kilogram is too large a unit, so that we take as our unit of mass the thousandth part of a kilogram, i. e. *a gram* [gr.], which may therefore be taken to be of the same mass as one cubic centimetre of distilled water at 4° C. The gram is also divided decimally as follows:—

Table of masses.

10 milligrams = 1 centigram.
10 centigrams = 1 decigram.
10 decigrams = 1 gram.
1000 grams = 1 kilogram.

Mass and Weight. It is necessary clearly to distinguish between these two terms. The *mass* of a body is the quantity of ponderable matter it is made up of. If we take a body from one place to another on the Earth's surface, or if we could take it to the Moon or carry it into the depths of space, our experience tells us that the quantity of matter it contains, i. e. its mass, is unaltered. The *weight* of a body, however, is the *force*

[2] The kilogram is very nearly equal to 2·2 lbs.

with which the Earth attracts it. This force of attraction is less the further we are from the centre of the Earth. At the top of a mountain a body weighs less than at sea-level. Again, since the Earth is flattened at the poles and bulges out at the Equator, the attraction of the Earth on the body is least at the Equator and increases as we travel towards the North or South pole. The weight of a body varies therefore according to its position upon and its height above the surface of the Earth[3].

We have no means of directly measuring the mass of a body since we are only cognizant of masses by the effect which forces have upon them. In an ordinary balance, when we weigh a body, we place it in one pan and add standard masses (so-called weights) in the other pan until the attraction of the Earth on the standard masses is balanced by the attraction on the body we are weighing. But, since the weight of a body is at a given place proportional to its mass, we may say that a certain body has a mass of 5 grams, when its weight balances the weight of a standard mass of 5 grams.

By the ordinary balance we cannot recognize that the weight of a body varies from place to place because the force of attraction due to the Earth is exerted both on the mass to be weighed and on the standard masses, so that, however this force varied, it would vary equally on the masses in the two scale pans. If, however, we weighed a given mass by a delicate spring balance in two different latitudes, we should then notice its difference in weight. It would be a profitable transaction to buy commodities on the Equator and sell them in higher latitudes, weighing them out in each case with a delicate Spring Balance!

We can no more lock up forces in a box than Pandora could imprison Hope in a casket, so that it is incorrect to talk of a *box of weights*—the correct term being a *box of masses*.

---

[3] Another cause of the variation of the weight of a body is the fact that the tendency of a body to fly off from the earth in consequence of its diurnal rotation varies according to the latitude.

B. Instruments.

1. The Linear Vernier.

A Linear Vernier is a short divided scale which can be made to move along the main scale, on which we are measuring, and by which we can measure to a greater degree of accuracy. A vernier may be divided in many ways, but the one usually met with is divided into ten equal divisions, its whole length being equal to nine main scale divisions, so that each division

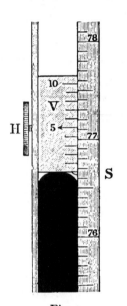

Fig. 1.

of the vernier is one-tenth shorter than a division of the scale. By its means we can accurately measure to a tenth of a scale division. To illustrate the method, Fig. 1 is a magnified diagram of the upper part of a barometer. S is the scale divided into centimetres and millimetres, V is the moveable vernier whose length is equal to nine millimetres and is divided into ten equal parts. It can be moved by the milled head H. The zero edge of the vernier has been brought so as to be exactly at the same height as the top of the mercury column, which is thus seen to be between 76·6 cm. and 76·7 cm. high. Looking along the vernier scale we notice that its 7th division is in the same straight line as one of the scale divisions, and since each vernier division is shorter than a scale division by $\frac{1}{10}$th of the latter, i.e. by ·1 mm., the distance between the

6th vernier division and the next scale division below it is ·1 mm.
5th   „   „   „   „   ·2 mm.
4th   „   „   ·3 mm.
3rd   „   „   ·4 mm.

. ˙ ˙2nd vernier division and the next scale division below it is ·5 mm.
1st   ,,   ·   ,,   ,,   ·· ,,   ·6 mm.
zero line of the vernier ,,   ,,   ,,   ·7 mm.

Therefore the true height of the mercury column is 76·67 cm.

Rules for using a Vernier. Bring the zero of the vernier exactly to the end of the length we are to measure : read off the scale division immediately below it. Looking along the vernier, notice the number of its division which coincides with a scale division. This number gives the tenths of a scale division by which the length to be measured is greater than the scale division above read.

2. The Sliding Callipers (Fig. 2).

. This instrument consists of a bar of metal *S*, having a scale of centimetres and millimetres engraved upon it. A jaw *A* is fixed at one end at right angles to it, and another parallel jaw *B*

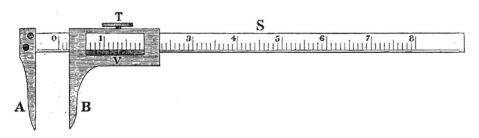

Fig. 2.

is fixed to a sliding piece on which a vernier *V* is etched, enabling the scale to be read to tenths of a millimetre. When the jaw *B* is pushed up flush with *A* the zero of the vernier ought to coincide exactly with the zero on the scale[4]. To make a measurement open the jaws, insert the bar, then push *B* up so that the two jaws touch the ends of the bar[5].

---

[4] This coincidence should always be tested before making a measure, and, if the zero marks do not coincide, a suitable correction must be applied to a reading.

[5] Be careful not to exert undue pressure, as it is easy to vary the reading by one or more small divisions by springing the frame of the instrument.

Now read off the length by the scale and the vernier. The screw *T* serves to clamp the vernier while the reading is being taken.

3. The micrometer screw-gauge (Fig. 3).

This instrument is used to measure the diameters of wires, thin sheets of metal, &c.; and consists of an accurately cut screw *S*, the distance between its threads, or its *pitch*, being

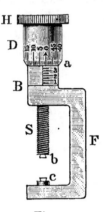

Fig. 3.

·5 mm. The screw works in the short metal block *B*. and is attached to the drum *D* which we can move backwards and forwards by the milled head *H*. The drum has its circumference at *a* divided into 50 equal parts. When the head *H* is turned the drum moves over the block *B*, along the length of which is engraved a straight line, divided into equal divisions, marked alternately by long and short lines. The distance between two successive lines, i.e. between a long and a short line, is ·5 mm.

The screw has a small plane tooth *b* which can be made to come into contact flush with another plane tooth *c*, attached to the supporting frame *F*. When the two teeth are in contact, the zero mark on the circular scale ought to be exactly in the same straight line as the linear scale on *B* at the zero division of *B*[6]. Ou turning the screw once round until the zero mark on the drum is again in the same straight line as the line on *B* but at a distance of one small division above where it started from, the screw has been moved through half a millimetre. Since the drum is divided into 50 equal divisions, on turning it until the next consecutive division on it is in the same straight line as the line on *B* we have moved the screw through $\frac{1}{50}$th of $\frac{1}{2}$ a mm., i.e. through ·01 mm. or ·001 cm. In order to measure with this instrument, e.g. to find the diameter of a given wire, separate the teeth, place the wire between them and turn the screw until the wire is just

[6] See Note 4.

pressed by both teeth [7]. On the linear scale on *B* read off the number of millimetres, and on the drum head read the number of that particular division which is in the same straight line as the line on *B*. This latter number gives the hundredths of a millimetre to be added to the number of millimetres just read on *B*, in order to get the required diameter of the wire.

Thus if the 37th division on the drum is in the same straight line as the line on *B*, the point of coincidence being between the 3rd and 4th division on *B*, we have

On the scale on *F*   3 small divisions = 1·5   mm.
On the drum   37   „   „   = ·37 mm.

∴   diameter of the given wire   = 1·87 mm. or ·187 cm.

4. The balance.

For a complete description of the balance we must refer the reader to a larger treatise, such as Stewart and Gee's *Practical Physics*, vol. I.   It will be sufficient here if we indicate what is meant by the sensibility of a balance, the methods usually adopted to increase the sensibility, to preserve the balance in good condition, and to use the instrument properly to estimate masses.   An ideal chemical balance of the ordinary pattern is essentially a lever of the first kind, i.e. a uniform rod or *beam*, balancing at the centre on a point called the *fulcrum*, the distances from the fulcrum to the ends of the beam, at which the pans of equal mass are suspended, being therefore equal. These distances are called the *arms* of the balance.   Generally the fulcrum and the points at which the pans are suspended are in the same straight line.   In order that the balance may be stable, its centre of gravity must be below the fulcrum.   If the centre of gravity coincides with the fulcrum, the balance would be in neutral equilibrium and would rest at any angle to the horizontal.   If the centre of gravity is above the fulcrum, the balance would be in unstable equilibrium, and the slightest motion of the beam would cause it to go down on one side.

----

[7] See Note 5.

The *sensibility* of a balance is measured by the angle through which it will turn for a given difference of masses in the pans.

Suppose AB is the beam, F the fulcrum, G the centre of gravity of the balance at which its weight W acts. Let a mass,

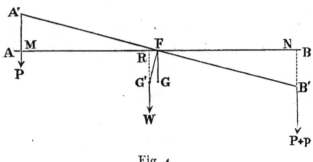

Fig. 4.

whose weight is P, be placed in one pan, and a mass, whose weight is $P+p$, in the other, and suppose in consequence it is deflected an angle AFA' or θ from its original horizontal position.

Let a be the length of the arms AF and BF, and l the distance FG of the centre of gravity below the fulcrum. Taking moments about F we have, for the condition of equilibrium in its displaced position,

$$P \cdot MF + W \cdot RF = (P+p)\,NF$$

or $\qquad P \cdot a \cos\theta + Wl \sin\theta = (P+p)\,a\cos\theta;$

$$\therefore \quad \tan\theta = \frac{a}{Wl}\,p.$$

Thus for a given difference p of the weights of the masses in the scale pans the angle θ or the sensibility will be greater

(i) The longer the arms (a) are.

(ii) The less the weight (W) of the balance.

(iii) The nearer the centre of gravity is to the fulcrum, i.e. the shorter l is.

The beam is made of brass in shape like an elongated lozenge with cross-beams, a form calculated to give rigidity combined with lightness (ii). At the centre of the beam there is a triangular steel prism with its knife edge turned downwards. At the ends of the arms are two other steel prisms with their knife edges turned upwards. When the balance is in use the central knife edge rests on an agate plate fixed at the top of the central brass pillar, and on each of the two end knife edges rests an agate plate, to which a scale pan is attached by a bent arm. To prevent friction of the bearings when the balance is not in use a moveable framework, called the Arrestment, attached to a brass tube, enclosing the central pillar, can be raised by a milled screw at the base, so that it raises the central knife edge off its agate plate and at the same time raises the agate plates, to which the scale pans are attached, off their knife edges. Above the central part of the beam is a bob of brass, called the gravity bob, which can be raised or lowered by a slow screw motion so as to raise or lower the centre of gravity of the beam, i. e. so as to make the term l less or greater (iii). The centre of the beam carries a long vertical pointer, behind the lower end of which is fixed a small graduated arc of ivory. The balance has a level attached to it and is enclosed in a glass case, the front of which can be raised at will to protect it and to prevent air currents when weighing.

The box of masses used with such a balance generally contain masses of the following gram-values

50, 20, 10, 10, 5, 2, 1, 1, ·5, ·2, ·1, ·1, ·05, ·02, ·01,

·01, ·005, ·002, ·002, ·001.

By combining together different masses we can get a range between 100 grs. and ·001 gr. The larger masses are made of brass, the decigrams and centigrams of Platinum and the milligrams of Aluminium. In using them always take them up by the small pair of pincers found in the box. If they are handled, the corrosion caused by the moisture of the hand alters their mass. The milligram masses are so small that they are liable to be lost, and it is not easy to make them with as great accuracy as the larger ones. We can, however, dispense with their use by the following means.

' Each arm of the balance is divided into nine parts[8], the graduations being numbered consecutively from the fulcrum outwards. Above the beam and slightly to one side of it there is a brass rod, which can be moved, parallel to the arm, from the outside by a milled head. This rod carries a small piece of bent wire which can be placed astride the arm. This piece of wire is called ' a rider' and has a mass usually of one centigram.

If we place the rider at the division marked 3, it would require $\frac{3}{10}$ths of its own mass, i. e. 3 milligrams, to be placed in the scale pan attached to the other arm to balance it. The effect on the balance arm of the rider placed at the third division is therefore the same as that of 3 milligrams in the pan. Similarly for other positions of the rider. So that, instead of adding milligram masses to the pan, we can place the rider at the proper division of the arm.

Precautions in weighing :—

1. See that the balance is levelled, and that the rider is in its place. If the pans are not clean, brush them with a flat camel's hair brush.

2. See whether the pointer swings equally on either side of the zero of the graduated arc. If not, a few grains of sand placed in one pan will cause it to do so.

3. Arrest the beam, place the body to be weighed in one pan. All substances liable to injure the pans must be placed in appropriate vessels: a watch glass of known mass will be found useful for the purpose.

4. In the other pan place the largest mass which is not too great and then, *without missing any*, add each smaller mass successively, removing it if too heavy, leaving it if too light.

5. Never add or remove a mass, or touch the balance while it is swinging, and never stop the balance with a jerk but only when the pointer is in its lowest position.

6. When you think you have added the proper mass, gently waft the air over one pan, close the window and placing yourself in a central position to avoid parallax, note whether the pointer

---

[8] In some balances the arms are divided into ten parts.

swings equally on either side of the zero. If it does you have the correct mass.

7. Get into the habit of reckoning up the masses from the vacant places in the box first; then check your result on replacing the masses in the box.

It is somewhat important to determine whether our assumptions that the scale pans are of equal mass and the arms of equal length are correct or not.

1. Suppose the arms of equal length: to compare the masses of the scale pans.

Let S and $S+\omega$ be the masses of the two scale pans, supposing them to be unequal. Place a body, such as a piece of glass, whose mass is X in the left-hand pan, weigh it and let its apparent mass be M grs.

then
$$S+X = M+S+\omega.$$

Interchange the scale pans and let the apparent mass of the body now be M',

then
$$S+\omega+X = M'+S.$$

Subtracting the two we get

$$\omega = \tfrac{1}{2}\,(M'-M),$$

which gives us the difference in the masses of the scale pans.

2. Suppose the scale pans of equal mass: to compare the lengths of the arms.

Place a body, such as a piece of glass, whose mass is X, in the left-hand pan, weigh it and let its apparent mass be M grs. If a is the length of the left arm, and b that of the right arm, taking moments about the fulcrum we have

$$Xa = Mb. \qquad (1)$$

Transfer the body to the right-hand pan and let its apparent mass now be M': we have

$$Xb = M'a. \qquad (2)$$

Dividing (1) by (2) we get

$$\frac{a}{b} = \frac{Mb}{M'a} \text{ or } \frac{a}{b} = \sqrt{\frac{M}{M'}},$$

which gives us the ratio of the lengths of the arms of the balance.

It is evident that by the double weighing we can find the mass of a body with a balance of unequal arms, for as above

$$Xa = Mb; \quad (1)$$
$$M'a = Xb. \quad (2)$$

Dividing (1) by (2) we get

$$\frac{X}{M'} = \frac{M}{X} \quad \text{or} \quad X = \sqrt{MM'}.$$

In order to get rid of any possible error due to the inequality of the arms of the balance we might perform the above double weighing in each case and take as the true mass the square root of the product of the two apparent masses, i. e. their geometric mean, as above. This 'method of double weighing' is usually ascribed to Gauss. Though in exceedingly accurate experiments this method is the best as the probable error of the result is least, it will be sufficient for our purpose to use the more expeditious 'method of substitution' due to Borda. The body to be weighed is counterpoised with shot and sand until the balance is in equilibrium [9]. The body is then removed and replaced by standard masses until equilibrium is again restored. These standard masses will evidently be equal to the mass of the body, since both they and the body, placed under the same circumstances, exactly counterbalance the counterpoise in the other pan, and thus any defect of the balance is neutralized. In some experiments it will be found necessary to counterpoise also a tare mass, but in either case, the original counterpoise is never altered during the course of the experiment.

Always state the result of a weighing in grams and decimals of a gram. Thus if a body was found to weigh 25 grams, 2 decigrams, 4 centigrams, and 7 milligrams, enter its mass as

25·247 grams.

As an exercise in the use of the balance, find out the mass of chalk required to write your name on the blackboard as follows. Counterpoise a piece of chalk and find its mass. Write your name on the blackboard with it five or six times. Weigh the

[9] The sand ought to be placed in a watch-glass, which also, of course, acts as part of the counterpoise.

chalk again and find its loss of mass. This loss, divided by the number of times you have written your name, will give you the answer required.

C. MEASUREMENT OF LENGTHS.

In measuring the distance between two points by a rule, place it on its edge and make one of the centimetre divisions coincide with one of the points. It is best not to measure from the end of the rule as it is never cut sufficiently sharp.

A more accurate result may be obtained, but using a pair of point compasses or 'dividers.' Apply the compasses so that its two feet coincide with the points whose distance apart we wish to measure, and then place them on the rule and so read off the length. Whichever method we adopt, we must always estimate to tenths of a millimetre. It is not difficult to imagine each millimetre divided into ten equal parts, and to estimate accurately that the length which is being measured is so many centimetres, so many millimetres and so many tenths of a millimetre. Enter your result in centimetres and decimals of a centimetre. Thus if you found a certain length to be 2 centimetres, 5 millimetres, and 3-tenths of a millimetre enter it as 2·53 cm.

N.B.—In every case when a number is written down as the result of an experiment be sure to add its denomination [cm., sq. cm., c.c., or gr. as the case may be]. Enter every single measure you make, and arrange your results in a tabular form wherever possible. As an example of the method see Experiment 10 [10].

1. To copy a scale on glass.

Apparatus. A rod made of hard wood, about 52 cm. long by 1 cm. wide—[near each end, perpendicular to its length, a stout needle is driven through the wood, each projecting about

[10] It will not be necessary to keep more than 3 decimal places. In order that the third place may be true, find the fourth decimal place. If this is less than 5 neglect it, but if it is 5 or a greater number add one to the third place.

2 cm. beyond the rod] : a Half-metre Rule : a Strip of glass, about 30 cm. long by 2 cm. broad : Solution of Hydrofluoric Acid : Paraffin-wax.

Experiment. Thoroughly clean the strip of glass by washing it successively with (1) Nitric Acid, (2) Distilled Water, (3) Caustic Soda, (4) Distilled Water, and dry it. When dry place a pellet of Paraffin-wax upon it, and holding it above a flame allow the melted wax to pass over the whole surface; then stand it up on its end to drain. When cold, it will be covered with a thin uniform layer of wax. Clamp it at its two ends near the edge óf the table, being careful to place a slice of cork between the clamp and the glass to prevent the latter breaking under the pressure. Now scratch on the glass with a needle two straight lines parallel to its length, one line [*A*] being ·5 cm., the other line [*B*] being 1 cm. from the edge nearest you. Clamp the ½-metre rule to the table in the same line with the glass strip, so that the adjacent ends of the rule and the glass strip are about 3 cm. apart. Place one needle of the rod on the line marked 10 cm. on the rule and with the other needle point scratch a line on the glass strip reaching from line *B* to the edge. Place the needle on each successive division of the rule and each time scratch a line on the glass strip, starting the centimetre lines from the line *B*, the millimetre lines from the line *A*, and every fifth millimetre from a point half way between the lines *A* and *B*. Having marked thus 25 cm. scratch numbers at the centimetre divisions in order from 1 to 25. Breathe on the scale, then brush the glass over with a solution of hydrofluoric acid, *being very careful not to get any of the acid on the fingers*, and leave it for ten minutes. On cleaning the paraffin-wax off with benzine, we see our scale etched on the glass. The divisions may be rendered more distinct by rubbing in rouge.

2. To find the ratio of the length of the circumference of a circle to its diameter.

Apparatus. (Fig. 5.) Circle Compass : Cardboard : Half-metre Rule.

Experiment. On the piece of cardboard draw three con-
centric circles. Measure the length of the circumference of one
of the circles with the Opisometer (Fig. 5). This consists of
a little frame fastened to a handle, and carrying a little wheel *A*,
armed with minute teeth, to prevent its
sliding over the paper without turning.
As the wheel turns, it at the same time
advances along the axis *BB'*, which is
fixed in the frame, and cut with a screw
thread of fine pitch. On proceeding to
make a measure with this instrument,
the wheel must start from the end *B*
of the screw. Place the pointer *B* at
a given point of the circumference and
carefully run the wheel along the whole
circumference back to the starting point.
Now run the wheel backwards along
the scale until it gets back to the end *B*
at which it started, when it will turn no
longer: the pointer at the end *B* then
shows on the scale the length of the
line that has been traversed. Repeat this measure twice more
and take the mean as the correct value. Measure the length
of the diameter. Do the same for the remaining two circles and
enter your results in a tabular form as follows:—

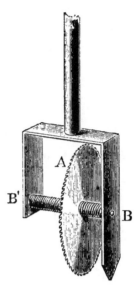

Fig. 5.

No. of Circle.	Circumference in cm. (l)	Diameter in cm. (d)	$\pi = \dfrac{l}{d}$	% error.
1	...	...	...	...
2	...	...	...	...
3	...	...	...	...

The numbers in the fourth column, obtained by dividing the
circumference of each circle by its diameter, ought to be constant.
If the experiment has been performed accurately we should
have found this constant equal to 3·1416 very nearly. This

ratio, which we shall indicate by the Greek letter π, may be taken roughly to be equal to $\frac{22}{7}$. The circumference of a circle therefore is equal to $\pi \times$ diameter, or $2\pi r$ cm., where r cm. is its radius.

Let us now find the percentage error of your results for π. Suppose the value of π deduced from Circle 1 was 3·137. This is ·005 too little. The percentage error is therefore

$$3\cdot137 : 100 :: \cdot005 : x,$$

$\therefore x = -\cdot16$, putting the negative sign because our value is too small. If our value is too large put $+$ sign. Find in this way the percentage error of your three results.

3. To measure the diameter of a sphere.

Apparatus. Two square-cut blocks of wood, whose thickness and width are greater than the radius of the sphere: a sphere [11] of about 3 cm. in diameter: Half-metre Rule: Callipers.

Experiment. The diameter of the sphere may be found by the following three methods :—

(*a*) Directly—by placing it between the two blocks of wood, which are pushed home so that one side of each presses against a plane surface (the wall), and another side of each touches the sphere between them. The perpendicular distance between two blocks gives us the diameter of the sphere. In measuring the distance be careful to hold the scale parallel to the outside edges of the two blocks. Repeat the measure twice more, altering the position of the sphere each time, and take the mean of the three readings as the correct value.

(*b*) Indirectly—by deducing the diameter from the length of the circumference, which is measured as follows. Make a mark on the sphere. Draw a straight line, longer than its circumference, on a piece of paper fixed to the table. Starting with the mark at one end of the line, roll the sphere carefully along it until the mark meets it again. The distance between the two points at which the mark touched the line gives us the length of the circumference of the sphere. Repeat this measure

[11]. A billiard ball will do.

twice more and take the mean of the three results as the correct value. Dividing it by π, we deduce the length of the diameter [Experiment 2].

(*c*) Directly—with the callipers. Measure the diameter by the callipers in two or three different places and take the mean as the correct value.

Compare the results you have obtained by the three methods. If there is a millimetre difference between them, these measures should be repeated.

Assuming the value last found is the true one, find the percentage errors of the other two.

4. To measure the diameter of a cylinder.

Apparatus. A Cylinder: Thread: Callipers: Half-metre Rule.

Experiment. The diameter of the cylinder may be found by the following two methods.

(*a*) Indirectly—wrap the thread round the cylinder five or six complete times. Take off the thread and measure the length that was wrapped round. Divide this length by the number of complete turns and we get the length of the circumference. Repeat this measure twice more, wrapping the thread round in each case a different number of times. Take the mean of the three results as the correct value. Dividing it by π, we get the length of the diameter.

(*b*) Directly—by the callipers. Measure the diameter by the callipers in two or three different places and take the mean as the correct value.

Compare the results you have obtained by the two methods. If there is a millimetre difference between them, these measures should be repeated.

D. Measurement of Areas.

Let *ABCD* be a rectangular parallelogram whose base *BC* is *b* cm., and whose height *AB* is *h* cm. in length. The area of the rectangle is *hb* sq. cm. A triangle has half the area of the

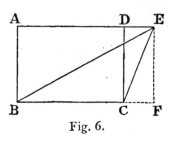

Fig. 6.

parallelogram on the same base and between the same parallels [Eu. i. 41]. Therefore the area of the triangle *BEC* is equal to $\frac{1}{2} AB \cdot BC$ sq. cm., i. e. $\frac{1}{2} BC \cdot EF$ sq. cm. Thus the area of any triangle is equal to half the product of its base into the perpendicular dropped from the opposite vertex on this base or the base produced. The length of this perpendicular, *EF*, is called the *altitude* of the triangle.

5. To find the area of a triangle.

Apparatus. Cardboard: Curve Paper[12]: Balance: Half-metre Rule: Scissors.

Experiment. Paste a piece of curve paper on the cardboard, and, when dry, draw with a fine pointed pencil as large a scalene[13] triangle as the cardboard will allow, and letter the vertices *A, B, C.*

The area may be found in the following three methods :—

(*a*) By reckoning up the number of square centimetres on the curve paper that the triangle encloses. The curve paper is divided up into square millimetres, and darker lines or lines of a different colour mark out square centimetres. Count up the number of complete square centimetres ·within the triangle. Then count up the number of square millimetres that complete

[12] See Appendix B. 2.
[13] To make sure the feet of the perpendiculars, to be drawn subsequently, may fall within the sheet of cardboard, it is best to draw an acute-angled triangle, though of course the same construction applies to all triangles.

the area of the triangle, neglecting round the periphery areas less than half a square millimetre, counting as whole ones areas greater than half a square millimetre. Remembering that 100 sq. mm. = 1 sq. cm., calculate the whole area of the triangle in square centimetres.

(*b*) Carefully draw perpendiculars from the points *A*, *B*, *C* respectively to the lines *BC*, *CA*, *AB* [Appendix B. 3] and letter the feet of the perpendiculars in order *D*, *E*, F^{14}. Measure the base *BC*, and the altitude *AD*. Multiply the two lengths together and take half the result. This gives you one value for the area of the triangle. Taking *AC* as your base, measure it, and also the corresponding altitude *BE*, and so get another value for the area of the triangle. Lastly, measure *AB* and *CF* and deduce a third value. Enter your results in a tabular form as follows :—

Length of		area in sq. cm. $(\frac{1}{2}bh)$
base in cm. (*b*)	altitude in cm. (*h*)	
$AB = \cdots$	$AD = \cdots$	...
$BC = \cdots$	$BE = \cdots$	...
$CA = \cdots$	$CF = \cdots$	...
Mean =		...

(*c*) The third method is by finding the mass of the triangle and dividing it by the mass of 1 sq. cm. of the cardboard used.

Cut out the triangle you have drawn, and counterpoise it on a balance. Remove the triangle and restore equilibrium by known masses, *M*, which gives us the mass of the triangle. From the same piece of cardboard cut out a rectangle. Find the area of the rectangle by multiplying together its length and

---

[14] If drawn correctly they ought to pass through the same point.

breadth [15]; suppose its area is a sq. cm.   Then counterpoise, weigh, and estimate its mass m.   Then

$\quad m$ grs. is the mass of a sq. cm. of the cardboard,

$$\perp \quad ,, \quad ,, \quad \frac{a}{m} \quad ,, \quad ,, \quad ,,$$

$$M \quad ,, \quad ,, \quad \frac{Ma}{m} \quad ,, \quad ,, \quad ,,$$

This gives us the area of the triangle.   Compare the results obtained in the three different methods.

Assuming the last result is the correct one, find the percentage error of the two others.

N.B.—By the first and third methods it is evident that we can determine the area of any irregular figure.

6. To find the area of a polygon.

Apparatus.   Same as in Experiment 5.

Experiment. Paste a piece of curve paper on the cardboard and, when dry, draw on it a polygon, say of four sides, and letter the vertices A, B, C, D.

(*a*) Determine its area in sq. cm. according to the method of Experiment 5 (*a*).

Length of		area in sq. cm. ($\frac{1}{2}bh$)
base in cm. (*b*)	altitude in cm. (*h*)	
$AC = ...$	...	...
$AD = ...$	...	...
$AE = ...$	...	...

Area of polygon =   ...

(*b*) Join the vertex A by straight lines to B, C, D, thus dividing

---

[15] It would be the easier plan to cut a rectangle out containing a definite number of sq. cm. as marked on the curve paper.

up the polygon into three triangles. Find the area of each triangle by measuring the base and its altitude as in Experiment 5 (*b*), and enter results in a tabular form as on preceding page.

(*c*) Find the area of the polygon by weighing according to Experiment 5 (*c*). Compare the results obtained.

Assuming the last result is the correct one, find the percentage errors of the others.

7. To prove by experiment that the area of a circle, whose radius is *r* cm., is πr^2 square centimetres.

Apparatus. Cardboard: Balance: Circle Compasses: Half-metre Rule.

Experiment. Describe a circle on a piece of cardboard: and through the centre draw two diameters at right angles to each other (see Appendix B. 4). At the four points at which these diameters meet the circle draw straight lines at right angles to them. By the intersection of these straight lines we have a square circumscribing the circle. Each of the small squares is the square on the radius and is of an area r^2 sq. cm. We wish to prove that the area of the circle is πr^2 or $3\frac{1}{7} r^2$ sq. cm., i. e. is equal to the area of three small squares and one-seventh of a small square.

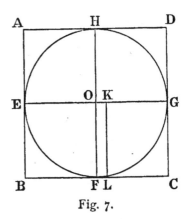

Fig. 7.

From *OG* mark off *OK* equal to $\frac{1}{7}$ of *OG* and draw *KL* parallel to *OF*, meeting *FC* at *L*. Cut out the figure *KGDABLK* and counterpoise on the balance. Now cut out the circle and put the pieces composing it back into the balance pan. It ought to remain exactly counterpoised, showing that the amount of card required to make up the circle is equal to that required to make up three and one-seventh of the squares on the radius.

8. To find the area of a circle.

Apparatus. Same as in Experiment 5 : Circle compass.

Experiment. Paste a piece of curve paper on the cardboard, and, when dry, describe with a fine pointed pencil as large a circle as the cardboard will allow. The area of the circle may be found in three different ways:—

(*a*) By reckoning up the squares on the curve paper enclosed by the circle in a similar way to Experiment 5 (*a*).

(*b*) Draw a diameter of the circle. Measure it carefully and calculate the radius, *r*. Deduce the area by multiplying the square of the radius by π. (Experiment 7.)

(*c*) By weighing the circle in a similar way to Experiment 5 (*c*). Compare your results by the three methods, and taking the last value as the correct one, calculate the percentage errors of the other two.

9. To find the area of the surface of a cylinder.

Apparatus. Cylinder : Half-metre Scale : Callipers.

Experiment. The area of the curved surface of a cylinder is equal to the product of the circumference of one end into the height. This is easily seen to be the case, for we can unfold a paper cylinder into a flat rectangle, the length of one side being the circumference, that of the adjacent side being the height of the cylinder.

Measure the height of the cylinder in two or three places and take the mean as the correct value, *h*.

Measure the diameter, *d*, of the cylinder by one of the two methods of Experiment 4. Multiply this length by π to get the length of the circumference, *l*: then the area of the curved surface is *hl* sq. cm.

From the diameter, measured above, find the radius, *r*, and hence calculate the area of the base of the cylinder, πr^2 sq. cm. The area of the surface of the cylinder is

$$(hl + 2\pi r^2) \text{ sq. cm.}$$

As an example of entering the results, the following results of an actual experiment are given.

To find the area of the surface of a cylinder.

Brass Cylinder A.

(*d*)		(*h*)
1·04 cm.		
1·03 cm.		
1·04 cm.		
$2 \mid \overline{1.036}$ cm. . . . . $\overline{3\cdot1416}$ 1·036 cm.		4·13 cm.
		4·12 cm.
$\therefore r = \cdot518$ cm.	$\overline{3\cdot255}$ cm. = circumference	4·13 cm.
$\cdot518$ cm.	4·127 cm.	4·127 cm.
$\therefore r^2 = \cdot268$ sq. cm.	13·433 sq. cm. (area of curved	
3·1416	surface)	
$\therefore \pi r^2 = \overline{.842}$ sq. cm.		
2		
$\overline{1\cdot684}$ sq. cm. . . . . 1·684 sq. cm. (area of two ends)		
$\overline{15\cdot117}$ sq. cm. = total area		

10. To find the area of a sphere.

Apparatus. Same as in Experiment 3.

Experiment. It can be proved by geometry that the area of the surface of a hemisphere is twice that of the circular base on which it rests. Therefore the area of the surface of a sphere is four times the area of a central section, or $4\pi r^2$ sq. cm. if the radius of the sphere is r cm.

Measure the diameter of the sphere in the three methods of Experiment 3 and take the mean of the results as the correct value. Hence deduce the radius. Square it and multiply it by 4π, which gives us the number of sq. cm. in the surface of the sphere.

11. To determine the sectional area and radius of a glass tube of fine cylindrical bore.

Apparatus. Two or three capillary tubes about 20 cm. in

length : Thermometer : Crucible : Balance : Half-metre Rule : clean Mercury.

Experiment. Clean the tubes by drawing through them successively (1) Nitric Acid, (2) Distilled Water, (3) solution of Caustic Soda, (4) Distilled Water [16]. Dry them carefully in an air bath. Counterpoise the empty crucible and a 5-gram mass on a pan of the balance. Draw up in one of the tubes a column of mercury. Measure its length in different parts of the tube. If your results differ, it shows the tube is not perfectly cylindrical. Take the mean of your results as the correct value, l cm. Transfer the column of mercury to the crucible, take away the 5-gr. mass and restore equilibrium by known masses. The difference between these and the 5 grams is evidently the mass, m, of the mercury column. Note the temperature, t, of the mercury and look in the tables [Appendix A. 2] for its density δ_t at this temperature. Since δ_t is the mass of 1 c.c. of mercury at this temperature, the volume of the mercury column is $\dfrac{m}{\delta_t}$ c.c.

But if r is the radius of the tube, the sectional area is πr^2 and the volume of the mercury column is $\pi r^2 l$ c.c. (Expt. 15) ;

$$\therefore \quad \pi r^2 l = \frac{m}{\delta_t},$$

or the sectional area is

$$\pi r^2 = \frac{m}{l \delta_t},$$

and the radius of the tube is

$$r = \sqrt{\frac{m}{\pi l \delta_t}}.$$

Repeat the same experiments for the other two tubes and enter your results in a tabular form as follows :—

[16] This may be done without soiling the hands by attaching a small glass syringe to the glass tube to be washed by half a metre of india-rubber tubing. Place one end of the glass tube in the liquid. By working the syringe we can cause the liquid to pass up and down the tube.

Tube	l in cm.	m in grs.	r in cm.	πr^2 in sq. cm.
1	...	...	...	...
2	...	...	...	...
3	...	...	...	...

The results above obtained will be required in Experiment 48.

E. Archimedes' Principle.

A solid when immersed in a fluid displaces a volume of the fluid equal to its own volume. If a solid is suspended in the fluid it apparently weighs less than when weighed in air, because the fluid exerts a resultant upward pressure or upthrust on the solid. Archimedes' principle is that this upthrust is equal to the weight of the displaced fluid, in other words a solid suspended in a fluid experiences a loss of weight equal to the weight of an equal volume of the fluid.

12. To prove Archimedes' principle by experiment.

Apparatus. A metal cube about 1 cm. inside: A metal box into which the cube exactly fits: Beaker: Thermometer: Balance: A wooden Ledge, which can be placed across a pan of the balance, of sufficient height not to touch it while a weighing is being made: Thermometer.

Experiment. (*a*) Suspend from one arm of the balance by a fine wire the metal cube so that it can be subsequently immersed in a beaker of water resting on the wooden ledge. Place the metal box on the pan and counterpoise box and suspended cube. Place a beaker of water on the ledge and allow the cube to dip completely in the water[17], and remove any adhering air-bubbles by a feather. Now fill the box with water by means

[17] Whenever we weigh a body suspended in water, be careful it does not touch the sides or bottom of the beaker.

of a pipette. We shall find that equilibrium is restored when the box is exactly filled, showing that the upthrust of the water on the cube, that is its apparent loss of weight, when suspended in water, is equal to the weight of an equal volume of water.

(*b*) To prove this by direct measurement, measure accurately the length, breadth, and height of the cube. Multiply the three together to get its volume, *v* c.c. suppose. Suspend it as before from one arm of the balance and counterpoise it. Place a beaker of water on the ledge so that the cube is completely immersed. Restore equilibrium by adding known masses, *m* grs. This represents the upthrust of the water on the cube, and we shall find that *m* and *v* are expressed by very nearly the same number. If we assume that 1 c.c. of water at the temperature of the experiment has a mass of 1 gram, this signifies that the upthrust on the body is equal to the weight of an equal volume of water.

This method of finding the upthrust on a body in water is in most cases sufficiently approximate to determine the volume of an irregularly shaped body. But a gram is the mass of 1 c.c. of water at $4°C$, therefore to get a more accurate value, note the temperature, *t*, of the water, and divide the mass, *m*, representing the upthrust, by the mass, Δ_t, of 1 c.c. of water (i. e. its density) at the temperature *t* as given in Appendix A. 1. The true volume of the cube above is therefore $\dfrac{m}{\Delta_t}$ c.c. Note what error is made by our assumption.

Repeat these experiments, using Turpentine in the place of water. In this case we must divide the upthrust by the density of Turpentine to get the true volume. Compare your results.

N. B.—Instead of the cube and its box we may use our cylinder for which we can make a waterproof case as follows:— Roll paper round the cylinder, and fill up the end with melted paraffin or plaster of Paris. The whole can be made waterproof by dipping in melted paraffin.

F. Measurement of Volumes.

13. To find the volume of a solid heavier than water by its displacement of water.

Apparatus. Large Stone: Measuring Jar: Pipette.

Experiment. A solid, immersed in water, displaces its own volume of water. Fill half the measuring jar with water. It will be found that, if the jar is of small diameter, the surface of the water will not be plane but concave. Add by means of the pipette water drop by drop, till a division mark is a tangent to the concave surface. Note the number of the division mark. Now introduce the stone into the jar and again read off the surface of the water. The difference of the two readings will be equal to the volume of the stone [18]. Repeat these observations twice more, altering the amount of water in the jar each time, and take the mean of your results as the correct value.

14. To prove that the volume of a sphere of radius r cm. is equal to $\frac{4}{3}\pi r^3$ c.c.

Apparatus. Same as in Experiment 3: Balance and Ledge: Thermometer: Beaker.

Experiment. Measure the diameter of the sphere according to the three methods of Experiment 3, and take the mean of your results as the correct value. Hence deduce its radius r, and find the value of $\frac{4}{3}\pi r^3$.

To prove that this is the volume of the sphere, we will find its upthrust in water. Suspend the sphere from an arm of the balance, so that it can subsequently be immersed in a beaker of water placed on the ledge. Counterpoise the sphere. Bring up a beaker of distilled water so that the sphere is completely immersed, removing any air-bubbles sticking to it by a feather. Restore equilibrium by known masses, m grs. This, according to Archimedes' principle, is the mass of an equal volume of water

---

[18] This would only be really the case if the jar had been graduated at the temperature at which the experiment is performed.

at the temperature *t*, which we observe with the thermometer. Look in the tables for the mass, Δ_t, of 1 c.c. of water at this temperature. Then $\dfrac{m}{\Delta_t}$ c.c. is the volume of the sphere. We shall find this to be the same number as we above calculated, thus proving that the volume of a sphere of radius *r* cm. is $\frac{4}{3}\pi r^3$ c.c.

15. To prove that the volume of a cylinder, whose height is *h* cm. and radius *r* cm., is equal to $\pi r^2 h$ c.c., i.e. to the area of its base multiplied by its height.

Apparatus. Same as in Experiment 4 : Balance and Ledge : Thermometer : Gas Jar : Measuring Jar : Beaker.

Experiment. Determine as in Experiment 4 the diameter of the base of the cylinder, and hence deduce its radius, *r*. The area of the base, πr^2, must be then calculated. Measure the height of the cylinder in two or three positions by the callipers and take the mean as the correct value *h*. Now calculate the value of $\pi r^2 h$ c.c.

To prove that this is the volume of the cylinder, find its upthrust in water just as in the case of the sphere (Experiment 14). If *m* is its upthrust in water, at the temperature, *t*, $\dfrac{m}{\Delta_t}$ c.c. is its volume, which we shall find to be the same number as we calculated above, thus proving that the volume of a cylinder is equal to $\pi r^2 h$ c.c.

As a further exercise find the volume of a cylindrical gas jar. Measure the inside height in two or three places and take the mean as the correct value *h*. Measure the inside diameter and hence calculate its volume $\pi r^2 h$ c.c. Check your result by filling it with water and then pouring the water into a measuring jar.

16. To prove that the volume of a sphere is equal to two-thirds of the volume of the circumscribing cylinder.

Apparatus. Same as in Experiment 3.

·*Experiment.* It is evident that the radius of the circumscribing cylinder is equal to the radius of the sphere and its height is equal to the diameter of the sphere (Fig. 8). Measure as before the diameter, h, of the sphere, deduce its radius r, and calculate the area πr^2 of the base of the cylinder. The volume of the cylinder is equal to $\pi r^2 h$ c.c.

Fig. 8.

Calculate the volume of the sphere $\frac{4}{3}\pi r^3$ c.c. This will be found to be equal to $\frac{2}{3}$ of $\pi r^2 h$, which proves what we require.

17. To find the thickness of a thin sheet of metal.

Apparatus. Half-metre Rule: Balance and Ledge: Beaker: Lead foil: Screw-gauge.

. *Experiment.* Smooth out all the creases in a piece of Lead foil and mark on it a rectangle about 10 cm. by 7 cm. Cut the rectangle out with a sharp knife and measure the two longer sides. Take the mean as the length, l. Measure the two shorter sides and take the mean as the breadth, b. The area of the rectangle is the product bl sq. cm. Call this area a sq. cm. Now find the volume of this rectangle by finding its upthrust in water as follows. Roll the rectangle into a cylinder and suspend it by a hair from an arm of the balance. Counterpoise it. Remove the sheet and restore equilibrium by known masses m_1 grs. This gives us its mass. Again, suspend it from the balance arm, and place a beaker of water on the ledge, so that the sheet is completely immersed. Read the temperature, t, of the water removing air-bubbles by a feather. Restore equilibrium by known masses, m_2 grs. It is evident that $m_2 - m_1$ grs., or m grs. suppose, is its upthrust in water. Therefore its volume in cubic centimetres is $\dfrac{m}{\Delta_t}$, where Δ_t is the mass of 1 c.c. of water at the temperature t of the experiment.

We can also determine its volume by dividing its mass m_1 by the density, i. e. the mass of 1 c.c., of Lead as given in Appendix A. 3. The mean of these two values must be taken for its true volume, v c.c.

Now   area × thickness = volume;

$$\therefore \text{ thickness} = \frac{\text{volume}}{\text{area}} = \frac{v}{a} \text{ cm.}$$

Measure its thickness directly by the screw-gauge and compare your results.

Find as above the thickness of some Copper Foil and Tin Foil.

18. To find the length of a twisted piece of Copper Wire.

Apparatus. A piece of twisted Copper wire of medium thickness (12 B. W. G.) : Screw-gauge : Balance and Ledge.

Experiment. Find the volume of the wire by finding its upthrust in water, and by dividing its mass by its density, in a similar way to that described in Experiment 17. Let v c.c. be the mean of the two values.

Measure its diameter in two or three places with the screw-gauge and take the mean of your results for its true diameter. Hence deduce its radius r and calculate its sectional area πr^2 sq. cm.

Now sectional area × length = volume;

$$\therefore \text{ length} = \frac{\text{volume}}{\text{sectional area}} = \frac{v}{\pi r^2} \text{ cm.}$$

In order to check your result measure off a certain length, l, of a straight piece of wire of the same material and gauge and weigh it. Let its mass be μ grs. Then the mass of 1 cm. of it is $\frac{\mu}{l}$ grs.

Find the mass, m_1, of the twisted piece of wire; then its length is evidently $m_1 \div \frac{\mu}{l}$ or $\frac{m_1 l}{\mu}$ cm.

Compare your results.

N. B.—It is evident that if we could measure its length l, we can, by determining as above the volume, find the radius

$$r = \sqrt{\frac{v}{\pi l}} \text{ cm.}$$

19. To graduate a glass vessel, i.e. to divide it into parts of known volume.

Apparatus.   A piece of stout glass 60 to 70 cm. in length and about 1 cm. in diameter : Clip : India-rubber tubing : Balance : Evaporating Dish : Paraffin-wax : Solution of Hydrofluoric Acid : Half-metre Rule : Thermometer.

Experiment.   Heat the glass tube near one end with the blow-pipe flame, turning it continually round so as to heat it uniformly.   When viscous, take it out of the flame and draw it out steadily.   When cool scratch the glass with a file at the contracted portion and snap it across.   Fuse the edges of the tube and fit on it tightly a piece of india-rubber tubing about 3 cm. long, provided with a clip.   Now cover the tube with a thin uniform layer of Paraffin-wax, and support it vertically in a clamp.   Counterpoise the evaporating dish and weigh out into it 10 grams of water.   Pour the water into the tube and with a needle make a horizontal scratch tangential to the curved surface of the water.   Add another 10 grams of water and do the same.   Continue this until the tube is nearly full. We have now the tube divided into a series of lengths approximately equal, if the tube is of uniform bore.   Divide each of these spaces into 10 equal divisions.   Now against the divisions scratch numbers, beginning with 0 at the top and marking every 5th division.   Brush some solution of Hydrofluoric Acid over the whole tube and leave it for ten minutes, so that the glass exposed by the scratches may be etched.   Melt off the paraffin, and rub in some rouge so as to make the divisions clearer.

It is now necessary for us to calibrate the tube, i. e. to determine the exact volume of each division.   Fill the burette with distilled water and note the temperature, t.   Into a previously counterpoised vessel, run water from 0 to the 5th division and weigh the amount of water to determine its mass m. If Δ_t is the mass of 1 c.c. of water at t° as given by the tables, $\dfrac{m}{\Delta_t}$ c.c. is the volume between the o and the 5th division. Dividing this volume by 5 we get the average volume of each division in this part of the tube. Repeat this, weighing the amount of water that runs through every five divisions, and, calculating its volume as above, enter your results in a tabular form as follows : —

(1) no. of divisions on burette.	(2) volume in c.c.	$\dfrac{(2)}{(1)}$ average volume per division in c.c.
0–5	...	
5–10	...	
...	...	
...	...	...

G. DENSITY OR SPECIFIC MASS.

The *density* or *specific mass* of a body at a certain temperature is the mass of a unit volume of it at this temperature. As bodies generally expand on being heated, the mass of a unit volume at a higher temperature is evidently less than it is at a lower temperature. The expansion of solids is so small however that we may neglect this variation of density within the range of temperatures we shall use. In the case of liquids and gases we must always specify the temperature to which the density refers.

If the mass of a body is M grs., and its volume V c.c., then its density D is $\dfrac{M}{V}$ grs. per c.c. or $M = VD$.

Density therefore is a concrete quantity. Thus the density of water at

$39 \cdot 2^\circ$ F or 4° C is $252 \cdot 769$ grains per cub. in.

or $62 \cdot 397$ lbs. per cub. ft.

or 1 gram per cub. cm.

The density of Lead at the same temperature

is 2856·2897 grains per cub. in.,

or 698·846 lbs. per cub. ft.,

or 11·2 grams per cub. cm.

Density therefore is expressed by different numbers according to the units we employ.

The *specific gravity* [20] of a body is the ratio of its mass to the mass of an equal volume of a standard substance at a standard temperature. In the case of solids and liquids, water at 4° C is taken as the standard substance, so that the specific gravity of a body is the number of times it is heavier than an equal volume of water at 4° C. It is therefore a ratio, or an abstract number, and is evidently the same number, whichever units we employ. Thus the specific gravity of Lead is 11·2 in all systems of units. In the metric system, where our unit of mass (1 gram) is the mass of the unit of volume (1 cub. cm.) of water at 4° C, both density and specific gravity are expressed by the same number, but we must remember it is concrete in the former, abstract in the latter case. To determine the density of a body, we have to divide its mass in grams by its volume in cub. cm. But its volume in cub. cm. is expressed by the same number as the mass in grams of an equal volume of water at 4° C, so that by the same experiment we get a result expressing the density of the body in grams per cub. cm. or its specific gravity.

In order to get a first approximation to the density of a body we may suppose that 1 c.c. of water at the temperature of the experiment has a mass of 1 gram. In this case the number of cub. cm. in the volume of the body represents the mass in grams of an equal volume of water. Therefore the density of a body is approximately found by dividing its mass by its upthrust in water.

i. *Of Solids.*

20. To find approximately the density of a large piece of mineral.

[20] A better name for this ratio is the ' relative density.'

Apparatus. Deflagrating Jar : Clip : Measuring Jar : Balance : a piece of a Mineral.

·*Experiment.* Fit a cork, through which passes a short piece of glass tubing, to a deflagrating jar, which must be supported, neck downwards, in a ring of a retort stand. On the glass tube fit a short piece of india-rubber tubing fitted with a clip. Scratch a horizontal mark·on the outside of the jar about 4 cm. from its wider end. Fill the jar with water up to the mark. Weigh the mineral and find its mass, M. Carefully lower it into the water until it rests at the bottom of the jar. The solid displaces its own volume of water which rises above the mark. Place a graduated jar underneath, open the clip, and let the water run out until its level returns to the mark. Read off the volume, V, of water that has run out. The approximate density of the body is therefore

$$\frac{M}{V} \text{ grams per cub. cm.}$$

In order to determine accurately the density of a body at $0°$ C, we should have to apply several corrections to the results of the weighings we have to make in the following experiments :—

(1) For the variation of the density of water. M grams of water only occupy M c.c. at $4°$ C. At the temperature, t, of the experiment their volume is $\dfrac{M}{\Delta_t}$ c.c., where Δ_t is the density of water at $t°$ as given in Appendix A. 1. The neglect of this correction can make an error of ·3 per cent. in the result. For a density of 20 the error would be ·08 in excess.

(2) For the upthrust of the air on the body weighed. By Archimedes' principle a body weighed in air apparently loses weight ·by an amount equal to the weight of the volume of air displaced. The greater the volume, i. e. the less the density of the body, the greater will be the error. The neglect of this correction can make an error of 2 units in the 2nd decimal place. For a density of 20 the error would be ·023.

(3) For the upthrust of the air on the standard masses, they only having their true values in vacuo. This correction is always very small, so that we shall neglect it.

The mass of air displaced in (2) and (3) depends on its density which varies with the temperature, pressure, and hygrometric state at the time of the experiment.

(4) For the expansion of the body and of the standard masses from o° to the temperature of the experiment. This affects their volume, which therefore affects (2) and (3). We shall neglect this error—in other words, the density we shall determine will be that at the temperature of the experiment.

(5) For the possible alteration of the temperature and pressure of the air between two weighings during our experiment. We shall suppose however neither to vary. Even if the temperature varied 5° and the pressure by 10^{mm}, the correction would only affect the 5th decimal place in the result.

(6) In very accurate experiments the mass of the thread by which we support the body from the balance arm ought to be allowed for as well as the upthrust on the part immersed when weighing in water. [See Note 22.]

In the following experiments we shall always apply correction No. (1), but, if we wish to neglect it, we have merely to put $\Delta_t = 1$ in the formulae. Appendix B. 1 will indicate how correction No. (2) is to be applied. If both corrections are applied, our result for the density at $t°$ ought to be true to the third decimal place and, if our assumption in (5) is valid, to the fourth decimal place. Give your result to two decimal places, working out to the third so as to get the second nearest the correct value. [See Note 10.]

21. To find the density of a solid heavier than water and not acted upon by water or by air.

Apparatus. Balance and Ledge: Piece of glass [21]: Thermometer: Beaker.

---

[21] A glass stopper free from air-bubbles will serve.

Experiment. Suspend the glass from the right arm of the balance by a hair [22] of such a length as to allow the glass to be immersed in a beaker of water, if placed on the ledge. Counterpoise the glass by shot and sand. Now remove the glass and restore equilibrium by placing known masses, M, in the left-hand pan. This gives us the mass of the glass. Bring up a beaker of distilled water on the ledge so that the glass, still suspended from the balance arm, is completely immersed in the water, being careful not to let it touch the sides or bottom of the beaker. Read the temperature, t, of the water. Remove air-bubbles by a feather and restore equilibrium by known masses, m. This evidently represents its upthrust in water, i. e. is the mass of an equal volume of water at $t°$. Look in the tables for the mass of 1 cub. cm. of water at $t°$. Suppose it is Δ_t. Then $\dfrac{m}{\Delta_t}$ is the volume of the glass in cub. cm. Therefore the density is

$$M \div \frac{m}{\Delta_t} \text{ or } \frac{M}{m} \Delta_t \text{ grs. per c.c.}$$

Find as above the density of Lead, Iron, Brass, Copper, Marble, &c., and enter your results in a tabular form as follows:—

Name of substance.	Mass in grams. M	Upthrust in water in grams. m	Temperature of the experiment. $t°$	Density of water at $t°$. Δ_t	Volume of body in c.c. $\dfrac{m}{\Delta_t} = v$	Density of body in grs. per c.c. $\dfrac{M}{v}$
...	...	...	...	...	...	...
		...	...	...	...	...
...	...	...	...	...	...	...

[22] Cocoon silk is a good suspension as, its density being very nearly the same as water, the upthrust on the portion immersed is practically nil.

22. To find the density of a solid lighter than water and not acted upon by water or by air.

Apparatus. Balance and Ledge: Beaker: a small heavy Sinker[23]: Beeswax: Thermometer.

Experiment. Attach the wax to the sinker and suspend them at the proper height above the ledge by a hair, as in Experiment 21, and counterpoise them. Then remove the wax and restore equilibrium by known masses, M, which gives us the mass of the wax. Attach the wax again to the sinker and bring a beaker of distilled water upon the ledge so that both are completely immersed, being careful not to let them touch the sides or bottom of the beaker. Remove any air-bubbles by a feather and read the temperature, t, of the water. Restore equilibrium by known masses, m_1. Remove the wax, the sinker still remaining in the water, and restore equilibrium by known masses, m_2. It is evident that $m_1 - m_2$ (or m suppose) represents the force tending to float the wax, that is, is equal to the excess of the mass of the water displaced by the wax over the mass of the wax itself. Therefore the mass of the water displaced, or of a volume of water equal to the volume of the wax, is $M + m$. If Δ_t is the mass of 1 c.c. of water at the temperature t as given in the tables, the volume of the wax is $\dfrac{M + m}{\Delta_t}$ and its density is

$$\frac{M}{M + m}\,\Delta_t \text{ grs. per c.c.}$$

Find as above the density of Paraffin-wax, Tallow, Cork, India-rubber, different kinds of wood, and enter your results in a tabular form similar to that given at the end of Experiment 21.

In many of the following experiments we shall require a 'density bottle.' It is made of glass and provided with an accurately ground stopper. A useful size is one capable of holding 50 c.c. of water. It is made of two patterns.

[23] A glass stopper may be used as a sinker, or, better, a small open cage of thick wire in which the solid can be put.

One, which we shall denote by a, has the stopper pierced with a very fine hole, through which the excess of liquid, when the bottle is filled, may be expelled when the stopper is pushed in tight (Fig. 9). The other pattern, which we shall denote by β, used generally when we have to deal with volatile liquids, has its stopper unpierced, but bears a hori-

zontal circle scratched on its stem. The essential value of each lies in our being able to fill it accurately with a definite volume of a liquid. We have, however, to take several precautions. In the first place, always place the bottle in water reaching up to its neck, so that we may be sure that the temperature at which it is filled succes-sively with different liquids during the course of an experiment is the same, and therefore that the volume of the bottle is unaltered. After filling a be careful no air-bubbles remain behind, and, on taking it out of the water by its neck do not handle it more than is absolutely necessary, as the heat of the hand would cause some of the liquid to overflow. After filling the bottle it is as well to take the stopper out and to place it and the unstoppered bottle on the balance-pan when weighing. The water in which the bottle is placed ought to have been standing near the balance for some time to attain the temperature of the air, otherwise the liquid is liable to expand on the bottle being taken out of the water. In filling β always arrange so that the circular mark on the stem is tangential to the curved surface of the liquid inside, adjusting the level with a pipette. Finally, with a piece of blotting paper, wipe off any moisture that may adhere to the inside of the neck above the mark.

Before using a density bottle see that it is clean and dry. To clean it, rinse it out successively with (1) Nitric Acid, (2) Distilled Water, (3) Caustic Potash, (4) Distilled Water, (5) Alcohol. To dry it, take a piece of hard glass tube about 20 cm. long and ·75 cm. in diameter. Draw out

Fig. 9

one end so that it may be sufficiently small to insert in the b ottle. Bend the tube at right angles about 5 cm. from this end. Fix the tube in a clamp with the bent end vertical and heat the horizontal portion with a large Bunsen. By means of india-rubber tubing attached to the larger end of the glass tube and to a pair of bellows, cause a gentle current of hot air to pass into the bottle, which must be placed resting mouth downwards on the upright portion of the glass tube. A very useful piece of apparatus for this purpose is the hand-bellows sold by chemists to be used with a throat spray.

23. To find the density of a solid in small pieces, not acted on by water or by air.

· *Apparatus.* Beaker : Thermometer : Balance : Leaden Shot: Density Bottle (*a*).

Experiment. Fill the clean density bottle with distilled water by immersing it completely in the water and push in the stopper. Take it out, and after drying it carefully, place it on the right-hand pan of the balance, as also a watch-glass containing some of the Leaden Shot [24]. Counterpoise them all. Remove the shot, being careful not to lose any, leaving the bottle and the watch-glass on the pan, and restore equilibrium by known masses, M. This gives us the mass of the shot. Take the bottle off the pan and put into it the shot, and, while completely immersed in water as before, push in the stopper. Be careful to get rid of all air-bubbles. Read the temperature, t, of the water. Take the bottle out, dry it, and replace it on the pan, and restore equilibrium by known masses, m. This evidently is the mass of water displaced by the shot. Hence, as before, the volume of the shot is $\dfrac{m}{\Delta_t}$, where Δ_t is the density of water at $t°$ as found in the tables. The density of the shot is therefore

$$M \div \frac{m}{\Delta_t} \text{ or } \frac{M}{m} \Delta_t \text{ grs. per c.c.}$$

Find as above the density of Glass, Lead, Copper, Marble, &c.

[24] We ought to take sufficient to fill about a third of the bottle.

in small pieces, and compare your results with those obtained in Experiment 21.

N.B.—In the case of a solid in small pieces soluble in water, we should have to use in the above method a liquid, of known density, *d*, which does not act on the solid.

*24. To find the density of a similar solid in the form of a powder [25].

Apparatus. Beaker: Thermometer: Balance: Density Bottle (*a*): Sand.

Experiment. Counterpoise the clean bottle along with a mass, μ, greater than the mass of the powder to be used plus the bottle full of water. Put into the bottle as much clean sand as will fill about a third of it. Take away the mass μ, replace the bottle on the pan and restore equilibrium by known masses, m_1. It is evident that $\mu - m_1$ grs. (*M* suppose) is the mass of the sand. Now fill up the bottle with distilled water [26], get rid of air-bubbles by the air-pump, if necessary, and push in the stopper as before. Read the temperature, *t*, of the water. Carefully dry the bottle and replace it on the balance pan, and restore equilibrium by known masses, m_2. Empty the bottle of its contents, rinse it out, and fill it with water as before. Dry it and replace it on the pan and restore equilibrium by known masses, m_3. We now have to find the volume of the sand. It is evident that $\mu - m_3$ grs. is the mass of water required to fill the bottle completely and $m_1 - m_2$ grs. the mass of water required to fill the bottle when it contains the sand. Therefore the mass of water displaced by the sand is $\mu - m_3 - (m_1 - m_2)$ grs., or m grs. suppose. Therefore the volume of the sand is $\dfrac{m}{\Delta_t}$ and its density is

$$\frac{\mu - m_1}{\mu - m_3 - (m_1 - m_2)} \Delta_t \quad \text{or} \quad \frac{M}{m} \Delta_t \text{ grs. per c.c.}$$

[25] This method is used for porous bodies, such as pumice-stone, when they are reduced to powder.

[26] If any of the powder floats on the top of the water, it must be collected on a watch-glass, dried, and weighed, and its mass must be subtracted from *M*.

Enter your results in a tabular form as follows:—

Tare mass (μ) = ... grs.

Mass added, 1st weighing (m_1) = ... grs.	Mass added, 3rd weighing (m_3) = ... grs.
$\therefore$ *mass M of sand* ($\mu - m_1$) = ... grs.	$\therefore$ mass of water completely filling the bottle ($\mu - m_3$) = ... grs.
$t = {}^\circ C : \Delta_t$ = ... grs. per c.c.	mass of water required to fill bottle with sand in it ($m_1 - m_2$) = ... grs.
$\therefore$ *volume of sand* $\left(\dfrac{M}{\Delta_t}\right)$ = ... c.c.	$\therefore$ mass m of an equal vol. of water = ... grs.

$\therefore$ density of sand $\left(\dfrac{M}{m}\Delta_t\right)$ = ... grs. per c c.

25. To find the density of a solid acted on by water but not by air.

Apparatus. Beaker : Thermometer : Balance and Ledge : Crystal of Copper Sulphate : Turpentine.

Experiment. Since crystallized Copper Sulphate is soluble in water, we cannot find its upthrust in water as in the previous experiments, but must use a liquid which has no action upon it, e. g. Turpentine. We must know or determine, as in a subsequent experiment, the density of the liquid. Suppose the known density of turpentine at the temperature of the experiment to be d_t. Suspend the crystal of Copper Sulphate, as before, by a hair from a balance arm, and counterpoise it. Remove the crystal and restore equilibrium by known masses, M. This gives us the mass of the crystal. Suspend the crystal in turpentine placed in a beaker, resting on the ledge, and, after removing air-bubbles, restore equilibrium by known masses, m. This gives us the upthrust in the turpentine, that is the mass of an equal volume of turpentine. But we know the mass of 1 c.c. of turpentine is d_t grams, therefore the volume of the crystal is $\dfrac{m}{d_t}$ c.c.   Thus its density is

$$\frac{M}{m} d_t \text{ grs. per c.c.}$$

Find as above the density of Alum [Turpentine], Rock-salt [Naphtha].

Now find the density of these substances in small pieces by the method of Experiment 23, but using the proper liquid instead of water.   Compare your results by the two methods.

26. To find the density of a solid acted on both by water and by air.

Apparatus.   Beaker : Thermometer : Balance : Crystallized Sodium Sulphate : Turpentine : Pipette : Density Bottle (β).

Experiment.   Place the clean bottle in a beaker of cold water reaching up to its neck, and fill it exactly up to the mark [27] with turpentine by means of a pipette.   Take it out of the water, dry it carefully, and counterpoise it along with a mass, μ, greater than the solid to be used.   Now place in the bottle a small clean crystal of Sodium Sulphate, being careful not to put in as large a piece as will cause the turpentine to overflow. Take away the tare mass, μ, and restore equilibrium by known masses, m_1.   Then $\mu - m_1$, or M, is the mass of the solid.   Place the bottle again in the beaker of cold water, and by a pipette withdraw sufficient turpentine so that its surface is again on a level with the mark.   After taking the bottle out and drying it, replace it on the balance pan and restore equilibrium by known masses, m_2.   Then $m_2 - m_1$, or m, gives the mass of turpentine displaced by the solid.   We must know or determine as in a subsequent experiment the density, d_t, of turpentine at the temperature of the experiment. Then, as before, the volume of the solid is $\dfrac{m_2 - m_1}{d_t}$ c.c. and its density is

$$\frac{\mu - m_1}{m_2 - m_1} d_t = \frac{M}{m} d_t \text{ grs. per c.c.}$$

Find as above the density of Sodium [Rock-oil].

[27] Refer to the method of filling as described on p. 38.

*27. To find the mass of a solid of known density imbedded in wax of known density.

Apparatus. A Leaden Bullet imbedded in a piece of Paraffin-wax. Same as in Experiment 21.

Experiment. Suppose the known density of the bullet is d_1, the known density of the wax is d_2. By proceeding exactly as in Experiment 21, we can find the mass m, volume v, and density d of the two together. We want to find the mass, m_1, of the bullet, and the mass, m_2, of the wax.

Now (1) $m = m_1 + m_2$,

(2) $m_1 = v_1 d_1$, $\Big\{$ where v_1, v_2, are the volumes of the

(3) $m_2 = v_2 d_2$, $\Big\{$ bullet and the wax respectively,

(4) $m = (v_1 + v_2) d$,

$\therefore$ (5) $m = v_1 d_1 + v_2 d_2$.

Treating (4) and (5) as simultaneous equations in the two unknowns v_1, v_2, we find by eliminating v_1 that

$$v_2 = \frac{m(d - d_1)}{dd_2 - dd_1}.$$

From (1)

$$m_1 = m - m_2 = m - v_2 d_2 = m - \frac{m(d - d_1) d_2}{dd_2 - dd_1};$$

$$\therefore m_1 = \frac{(d_2 - d) d_1}{(d_2 - d_1) d} m.$$

Substituting the known numbers for d, d_1, d_2 and m, we get the mass of the bullet m_1. Similarly for mass of wax.

If we knew the mass of the bullet we could by the same process find its density. In a similar method to the one described Mr. Joly determined the density of minute specimens by imbedding them in paraffin-wax. It is easy to deduce that the density d_1, of a minute specimen of known mass m_1, is, using the same letters as before,

$$d_1 = \frac{m_1 dd_2}{md_2 - (m - m_1) d}.$$

*28· To determine the thickness of a hollow sphere of metal of known density.

Apparatus. Balance : Hollow Metal Sphere : Callipers.

Experiment. Suppose the known density of the metal, of which the hollow sphere is made, is d grs. per c.c. Measure with callipers the diameter of the sphere and hence deduce its radius R. The volume of the sphere is then calculated from $\frac{4}{3}\pi R^3$ to be V c.c., suppose. If the sphere were solid, its mass would be Vd grs. By the balance find the mass, m, of the metal shell. The volume of the shell is therefore $\frac{m}{d}$ c.c. Hence the volume of the hollow core is $V - \frac{m}{d}$, (v c.c. suppose). If x is the radius of the core, we can find its value because $v = \frac{4}{3}\pi x^3$. Knowing R and x we can determine $R - x$ the thickness of the shell.

ii. *Of Liquids.*

29. To find the density of a liquid by the density bottle.

Apparatus. Beaker : Thermometer : Balance : Mercury : Density Bottle (a).

Experiment. Counterpoise the clean bottle along with a mass, μ, greater than the mass of liquid that fills the bottle. Place the bottle in a beaker of cold water reaching up to the neck. Fill it completely with clean mercury and push the stopper in. Take the bottle out and dry it, being careful not to handle it more than is absolutely necessary, and replace it in the balance pan. Take away the tare-mass, μ, and restore equilibrium by known masses, m_1. Then $\mu - m_1$, or M, gives us the mass of mercury filling the bottle. Empty out the mercury, and place the bottle again in the beaker of cold water and fill it with distilled water. Take it out, dry it, and replace it in the balance pan and restore equilibrium by known masses, m_y. It is evident that $\mu - m_2$ or m gives us the mass of water filling the bottle. If t is the temperature of the water and Δ_t the density of water at this temperature, the volume of the water as well as of the mercury is $\dfrac{m}{\Delta_t}$, hence the density required is

$$\frac{\mu - m_1}{\mu - m_2} \Delta_t = \frac{M}{m} \Delta_t \text{ grs. per c.c.}$$

Find as above the density of Turpentine, Oil, Glycerine, Alcohol, 15 % solutions of Copper Sulphate, and of Salt, and other liquids; entering your results in a tabular form similar to the one shown in Experiment 21.

N.B.—In the determination of the density of volatile liquids it is preferable to use the density bottle β.

30. To find the density of a liquid by finding its upthrust on a solid.

Apparatus. Two Beakers: Thermometer: Balance and Ledge: Glass-stopper: Turpentine.

Experiment. Suspend the glass-stopper in a similar way as described in Experiment 21. Counterpoise it: bring up on to ledge a beaker containing water so that the stopper is completely immersed. Remove air-bubbles and restore equilibrium by known masses, m. This represents the upthrust on the stopper due to the water, and is the mass of a volume of water equal to the volume of the stopper. If t is the temperature of the water and Δ_t its density, $\dfrac{m}{\Delta_t}$ c.c. is the volume of the stopper. Dry the stopper and, replacing the beaker of water by one containing turpentine, find in a similar way the upthrust due to the turpentine. Suppose it is M grs. This is the mass of an equal volume of turpentine, and the density required is

$$\frac{M}{m} \Delta_t \text{ grs. per c.c.}$$

Find as above [28] the densities of all the same liquids, except mercury, as were used in Experiment 29, and compare your results.

N.B.—This method is well adapted for very volatile liquids.

---

[28] In case of acids we must use a suspension which is not acted upon by them, e g. fine Platinum wire.

31. To find the density of a liquid by means of a U-tube [29].

Apparatus. A *U*-tube with legs about 50 cm. long (Fig. 10):
Beaker: Thermometer: Half-metre Rule: clean dry Mercury.

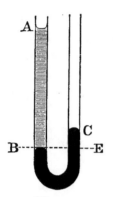

Fig. 10.

Experiment. Support the *U*-tube vertically, using a plumb-line. Pour into it some clean dry mercury until its level is about 12 cm. above the base. Now nearly fill one arm with pure distilled water, and suppose the final levels of the liquids are as in Fig. 10, where *AB* is the water, and *BC* the mercury column. Since the liquids are in equilibrium the pressure on the surface of separation at *B* is equal to the pressure in the same horizontal line at *E*.

If the area of the tube is a,

   ,,   height of the water $AB = h$, and its density at $t°$, Δ_t,

   ,,   height of the mercury $CE = h'$,,   ,, d_t,

the pressure due to the mass $ha\Delta_t$ grs. of water is equal to

                    ,, $h'ad_t$ grs. of mercury.

$$\therefore \; ha\Delta_t = h'ad_t \;\; \text{or} \;\; \frac{d_t}{\Delta_t} = \frac{h}{h'}.$$

Thus the densities of two liquids in equilibrium in a *U*-tube are inversely proportional to their heights above their common surface.

Measure [30] the height of *B* above the bench and subtract it from the height of *A* above the bench. This gives us *AB* or h. Now measure the height of *C* above the bench and also subtract from it the height of *B* above the bench. This gives us *CE* or h'.

---

[29] This method must be only used with liquids that do not mix with each other.

[30] The surface of a liquid in a tube is not in general plane but curved, the more so as the tube is narrower. When reading heights always read to the lowest or highest point of the curve, according as it is concave or convex. A piece of white paper placed behind the tube will facilitate the reading of the levels.

Looking in the tables for Δ_t, we get the density of mercury

$$d_t = \frac{h}{h'}\,\Delta_t \text{ grs. per c.c.}$$

Determine the density two or three times more, altering the levels each time by adding more liquid, and take the mean as the correct density.

Compare as above the densities [31] of Oil and Turpentine each with mercury : and then knowing the density of mercury find them referred to water. Then find their densities directly with water and compare your results.

32. To find the density of a liquid by the W-tube [32].

Apparatus. (Fig. 11.) Half-metre Scale : Thermometer : Glycerine.

Experiment. ABB', CEE', are two U-tubes. The ⁺ two shorter legs are joined tightly together with a piece of india-rubber tubing I. Glycerine is poured into the tube AB, and water into CE. Air is enclosed thus between B' and E', and on alternately adding more of the liquids the air is compressed in $B'E'$. Be careful not to add as much as will force either liquid into the horizontal arm. After one or two repetitions of the operation the liquids are made to arrange themselves as in the figure. The pressure at $B' =$ the pressure at E', since the air in $B'E'$ is all at a constant pressure.

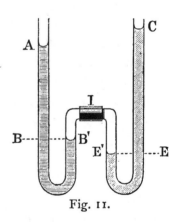

Fig. 11.

Suppose the height of the top of the glycerine at A above B', (i. e. AB) $= h$ and the density of glycerine d_t; and the height of the top of the water at C above E', (i.e. CE) $= h'$ and the density of water Δ_t.

---

[31] Always pour in the heavier fluid first and do not pour in so much of the lighter as will cause the common surface to be forced round the bend.

[32] This method may be used whatever the liquids, but must be used with liquids that would mix together.

If a is the sectional area of the tubes, the pressure at B' or at B is $ha d_t$, the pressure at E' or E is $h'a\Delta_t$ gram weight;

$$\therefore \quad ha d_t = h'a\Delta_t;$$

$$\therefore \quad d_t = \frac{h'}{h}\Delta_t \text{ grs. per c.c.}$$

The heights are to be measured from the bench as explained in Experiment 31.

Determine the density two or three times more, altering the levels each time by adding more liquid and take the mean of your results as the correct value.

Find as above the densities of the same liquids as used in Experiment 29, and compare your results.

N.B.—Instead of using a W-tube we may proceed as follows: Take two clean glass tubes of same bore about 80 cm. long. Join one end of each tube by an india-rubber T-tube, and arrange them vertically with their other ends dipping, one into a beaker of water, the other into the liquid whose density we require. Suck the liquids up into the tubes by means of the third arm of the T-tube, then clip this arm. Adjust the levels of the liquids in the beakers to the same height. The densities of the two liquids are inversely proportional to the vertical heights of the columns above their surface in the beakers, which are measured as above.

iii. *Of Gases.*

*33. To determine the density of air [33].

Apparatus. Balance : Thermometer: Bunsen Burner : two round bottom flasks of about 250 c.c. as nearly as possible of the same size [34] [each flask must be fitted with a cork through which passes a piece of glass tubing about 3 cm. long, which has a piece of india-rubber tubing about 3 cm. long fitted on it]: a Clip.

Experiment. Be very careful to see that the cork and tubings of one of the flasks are fitted quite air-tight. Support this flask

---

[33] Ramsay's Chemical Theory.
[34] Two flasks are used in order to eliminate the error due to the buoyancy of the air.

by a wire round its neck to the hook at the end of the right-hand arm of the balance. Support the other flask, fitted as above, to the hook on the left-hand arm. Add a tare mass μ, of about 50 grs., to the right-hand pan, and add a counterpoise to the left-hand pan till the balance is in equilibrium. Take the right-hand flask off the hook and pour about 30 c.c. of distilled water into it. Take the clip off, and fit the cork tightly into the flask. Boil the water in the flask, taking care not to use too large a flame, else the glass, not in contact with the water, will be heated, and will crack when the water touches it. Let the steam blow out for about five minutes. Close the india-rubber tubing with the clip and *immediately* remove the flame. Wipe the flask, allow it to cool, replace it on the hook, take away the tare mass, and add masses m_1 to restore equilibrium. The air that was in the flask has been expelled by the steam and the flask contains only the water and its vapour. Now find the temperature t of the water in the flask and find Δ_t, its density from the tables. The mass of water in the flask is $\mu - m_1$ and its volume is

$$\frac{\mu - m_1}{\Delta_t} \quad \text{or} \quad V' \text{ c.c., suppose.}$$

Open the clip for a few seconds to let the air in, and replacing the flask on the hook restore equilibrium by known masses, m_2. The increase of mass $m_1 - m_2$ evidently is the mass of air that has entered.

Now we have to find the volume of the flask. Fill it up to where it was clipped with distilled water and find the mass, M, of the water. The volume of the flask is $\dfrac{M}{\Delta_t}$, or V c.c. suppose. The volume of the air that entered therefore is $V - V'$, or v c.c. suppose. This air is at a temperature t and a pressure $H - F$, where H is the barometric height, as given by the barometer, and F the maximum pressure of aqueous vapour [35] at the temperature, t, of the water, which we find from the tables. Therefore reducing to standard temperature and pressure, this

---

[35] See *Practical Work in Heat* by the author.

E

volume, v, becomes by Boyle and Charles' law,

$$\frac{(H-F)\,v}{76\,(1+at)}\ \text{c.c. at o}^\circ\text{ C and under 76 cm. pressure,}$$

where $a = \cdot00366$, the coefficient of expansion of air.

Its mass we found to be $m_1 - m_2$ grs., therefore the density of air, or the mass of 1 c.c. at o$^\circ$ C and 76 cm., is

$$(m_1 - m_2) \div \frac{(H-F)\,v}{76\,(1+at)}\ \text{grs.}$$

iv. *Hydrometers.*

34. To find the density of a solid, not acted upon by water or by air, by means of a Hydrometer of constant immersion.

Apparatus. Nicholson's Hydrometer: Beaker: Thermometer: Box of Masses : Wide tall glass Jar : Piece of glass.

Experiment. A Nicholson's Hydrometer consists of a hollow cylindrical metal body, able to float in water. Its ends are closed by conical surfaces to prevent the instrument sinking too rapidly if overweighted. It bears a stem, which has a mark scratched upon it, at the top of which is a metal cup. It also has a metal cup at the bottom. Partly fill the glass jar with distilled water, so that the Hydrometer may be immersed up to the mark on the stem without touching the sides or the bottom of the jar. Float the Hydrometer in the water and remove all air bubbles adhering to it. Place masses, m_1, in the upper pan until it sinks exactly to the mark on the stem[36]. It is somewhat hard to observe this with accuracy, so that the best way is to cause the instrument to oscillate up and down. If the mark on the stem moves an equal distance above and below the surface of the water the masses added are correct. Remove the mass m_1, place the piece of glass[37] in the pan, and add masses, m_2, till the Hydrometer again sinks to the mark.

[36] To prevent the instrument sinking too far through overloading, a wire fork through which the stem passes should always be placed across the top of the jar.

[37] The solid must not be so heavy as, of itself, to cause the Hydrometer to sink below the mark on the stem.

Then $m_1 - m_2$ is evidently the mass of the glass. Transfer the glass to the lower pan, remove any air bubbles on it and add masses, m_3, to the upper pan till the Hydrometer again sinks to the mark. Then $m_3 - m_2$ represents the upthrust of the water on the glass, i. e. is equal to the mass of an equal volume of water. Read the temperature, t, of the water and find the value of Δ_t from the tables. The volume of the glass is $\dfrac{m_3 - m_2}{\Delta_t}$ c.c. and its density is

$$m_1 - m_2 \div \frac{m_3 - m_2}{\Delta_t} \quad \text{or} \quad \frac{m_1 - m_2}{m_3 - m_2} \Delta_t \text{ grs. per c.c.}$$

Find as above the densities of the solids used in Experiment 21 and compare your results.

N.B.—In the case of a solid lighter than water, when it is placed in the lower pan, it must be fastened down with a piece of wire fixed to the instrument. This wire of course must be attached to the instrument at the beginning of the Experiment.

Find the densities of the solids used in Experiment 22 and compare your results.

35. To find the density of a liquid by a Hydrometer of constant immersion [38].

Apparatus. Nicholson's Hydrometer: Box of Masses: Thermometer: Tall wide glass Jar: 15% Solution of Copper Sulphate.

Experiment. Find the mass, M, of the Hydrometer in the usual way. Nearly fill the jar with the solution of copper sulphate. Add masses, m_1, to the upper pan till the instrument sinks to the mark. Replace the copper sulphate by distilled water and find the mass, m_2, to sink the instrument in the water up to the mark. Find temperature, t, of the water and Δ_t from the tables. In both cases the same volume of liquid was displaced. The mass of this volume of copper sulphate is equal to $M + m_1$ grs., and the mass of the equal volume of

---

[38] This method is only available in the case of liquids sufficiently dense to prevent the Hydrometer itself sinking below the mark, and must not be used in the case of acids or other liquids which would attack the instrument.

water is $M + m_2$ grs. The volume of the liquids displaced is

$$\frac{M + m_2}{\Delta_t} \text{ c.c.,}$$

therefore the density of copper sulphate is

$$\frac{M + m_1}{M + m_2} \Delta_t \text{ grs. per c.c.}$$

Find as above the density of a 15% solution of common salt and Glycerine: and compare your results with those obtained in previous Experiments.

36. To find the density of a liquid by a Hydrometer of variable immersion.

Apparatus. A Hydrometer made as described below: Beaker: Thermometer: Half-metre Rule: Turpentine.

Experiment. Take a glass tube about 10 cm. long and 1·5 cm. in diameter, closed at one end [a test-tube will do]. Gum a narrow strip of paper along its length and graduate it by pouring in one c.c. of water [39] at a time, and with a pencil mark on the paper the levels of the liquid. Take a similar strip of paper, and carefully mark off on it divisions equal to those just made, and number them from o upwards. Gum it inside the tube with the divisions facing the glass, the zero being at the bottom, so that they are on the same level with those on the outside strip. Wash off this latter strip and pour as much mercury into the tube as will make it float in water with about half its volume immersed.

When a body floats in a liquid in equilibrium, its weight, acting downwards, must be equal to the upthrust of the liquid on the part immersed, i. e. must be equal to the weight of the liquid displaced, by Archimedes' principle. Float the Hydrometer in water and read the division on the scale where the water level cuts it. Suppose this is at the n_1th division. After drying the Hydrometer let it float in turpentine, and suppose the level of the turpentine cuts it at the n_2th division. The mass of the Hydrometer is equal both to that of n_1 c.c. of

---

[39] Or equal volumes not necessarily of one c.c.

water, and to that of n_2 c.c. of Turpentine at the temperature t, which we must observe. If Δ_t is the density of water at this temperature, the mass of

$$n_2 \text{ c.c. of Turpentine} = \text{mass of } n_1 \text{ c.c. of water at } t^\circ$$
$$= n_1 \Delta_t \text{ grs.;}$$

therefore the mass of

$$\text{1 c.c. of Turpentine} = \frac{n_1}{n_2} \Delta_t \text{ grs.}$$

Repeat this twice more, adding a little more mercury to the Hydrometer each time, and take the mean of your results as the correct value of the density.

Find as above the density of a solution of Copper Sulphate, Salt Solution, Alcohol, Glycerine, and compare your results with those obtained before.

Having now determined the densities of various solids and liquids by different methods, it would be well if the student collected his results in a table—one table being made for solids, another for liquids. Comparing them with the values given in the Appendix, notice which methods give the best results.

v. *Contraction in mixtures.*

* 37. **To determine the contraction which takes place when a solid is dissolved in water.**

Apparatus. Corked flask of about 250 c.c. capacity : Beaker : Thermometer : Balance : Crystallized Sodium Sulphate : Pipette : Density Bottle β : Thermometer : Turpentine.

Experiment. First determine the density, d, of crystallized Sodium Sulphate as in Experiment 26. Counterpoise the corked flask along with a mass μ (about ninety grams) greater than the mass of the solution you intend to use. Place in the flask about 10 c.c. of the salt and restore equilibrium by known masses, m_1. The mass of the salt added is therefore $\mu - m_1$ grs. and its volume $\frac{\mu - m_1}{d}$, or v_1 c.c. suppose. Now add about 100 c.c. of distilled water, and allow the salt to dissolve entirely. Replace the flask on the pan and restore equilibrium by known masses, m_2.

Allow the solution to remain for a few minutes so that it may attain the temperature of the room. Then take the temperature, t, of the solution and find the density of water Δ_t. The mass of water added is $m_1 - m_2$ grs. and its volume is

$$\frac{m_1 - m_2}{\Delta_t}, \text{ or } v_2 \text{ c.c. suppose.}$$

Determine the density D of the solution as in Experiment 29. The mass of the solution is $\mu - m_2$ grs. and its volume is therefore

$$\frac{\mu - m_2}{D}, \text{ or } v \text{ c.c. suppose.}$$

If no contraction had taken place, v would be equal to $v_1 + v_2$, therefore the contraction, when a volume $v_1 + v_2$ c.c. of the solution is made, is $v_1 + v_2 - v$ c.c.; $\therefore$ on making 100 c.c. of the solution the contraction is

$$\frac{v_1 + v_2 - v}{v_1 + v_2} \times 100 \text{ c.c.}$$

Again, in $\mu - m_2$ grs. of the solution there are $\mu - m_1$ grs. of the salt, and therefore there is a contraction $v_1 + v_2 - v$ c.c. when the solution contains

$$\frac{\mu - m_1}{\mu - m_2} \times 100 \text{ per cent. by mass of the salt.}$$

Enter your results in a tabular form as follows:—

	Mass in grs.	Density in grs. per c.c.	Volume in c.c.	
Salt . .	...	...	...	$\mu = \ldots$ grs.
Water .	...	...	...	$m_1 = \ldots$ grs.
Solution	...	...	...	$m_2 = \ldots$ grs.
				$t = \ldots$ C

$\therefore$ contraction =   ...   % of salt present = ...

Repeat the above, taking different proportions of the salt and water. We may lessen the labour of making a fresh solution as follows:—Replace the flask with its contents on the balance pan and restore equilibrium by known masses, m_3. The mass,

$\mu - m_3$, of the solution remaining consists of salt and water in the ratio $\mu - m_1 : m_1 - m_2$;

$\therefore$ the mass of salt in it is $\dfrac{\mu - m_1}{\mu - m_2}(\mu - m_3)$ grs.,

and mass of water in it is $\dfrac{m_1 - m_2}{\mu - m_2}(\mu - m_3)$ grs.

Add water or salt [40] again according as a weaker or stronger solution is required and find the mass of either added as before. This gives us the data for finding the total mass of water and salt in the solution. Determine its density, and, after calculating the contraction as above, construct another table. If you are able to determine a sufficient number of results, draw a curve [41] showing the relation between the percentages of salt present and the corresponding contractions at $t°$.

* 38. To find the percentage of Absolute Alcohol in the Alcohol in use in the Laboratory.

Apparatus. Same as in Experiment 29.

Experiment. Determine the density of the Alcohol in use as in Experiment 29 and note its temperature. We can find the percentage of Absolute Alcohol by means of the table in Appendix A. 5. To explain how this is done we will take the result of an experiment by which it was found that a certain sample of Alcohol had a density of ·805 grs. per c.c. at the temperature of 17° C. From the table we find that a mixture containing 95% of Absolute Alcohol for a rise of temperature from 15° to 20° C has its density decreased by ·80864 − ·79553 = ·00429, $\therefore$ its density at 17° would be $\frac{2}{5} \times$ ·00429 or ·00172 less than that at 15°, i.e. its density would be ·80692. In a similar way we find the density of Absolute Alcohol at 17° C is ·79199. Thus at 17° a decrease of 5% of Absolute Alcohol causes the density to increase by

---

[40] Remember that at

10° 100 grs. of water can dissolve about 10 grs. of sodium sulphate

15° ,,   ,,   ,, 15 grs. ,, ,,

20° ,,   ,,   ,, 20 grs. ,, ,,

[41] See Appendix B 2.

·80692 — ·79199 = ·01493. The density of the Alcohol in use
is ·80692 — ·805 or ·00192 less than the 95% solution. If an
increase of 5% causes the density to decrease by ·01493, what is
the increase % corresponding to a decrease of density ·00192?

The answer is $\dfrac{5 \times ·00192}{·01493}$ or ·65 nearly.

Therefore the solution used contains 95·65% of Absolute
Alcohol.

***39. To determine the contraction which takes place
when Alcohol is mixed with water.**

Apparatus. Same as in Experiment 29: Density Bottle β.

Experiment. Determine as in Experiment 29 the density, d,
of the Alcohol, noting its temperature, $t°$, carefully. Counterpoise the density bottle along with a mass μ (say 60 grs.)
greater than the mass of the mixture to be used. Place the
bottle in a beaker of water reaching to the neck and half fill it
with distilled water, which ought to have been standing near
the balance for some time to attain the temperature of the air.
Take the bottle out, dry it carefully, and place it on the balance
pan. Remove the tare mass μ, and restore equilibrium by
known masses, m_1. Then $\mu - m_1$ is the mass of the water, and if
Δ_t is its density at $t°$ its volume is

$$\frac{\mu - m_1}{\Delta_t}, \text{ or } v_1 \text{ c.c., suppose.}$$

Replace the bottle in the beaker of water, and nearly fill it
with alcohol. Since heat is generated, let it stand in the water
for a few minutes until it cools to its original temperature.
Take it out, dry it, and replace it on the scale pan and restore
equilibrium by known masses, m_2. The mass of Alcohol added is
$m_1 - m_2$ grs. and its volume is

$$\frac{m_1 - m_2}{d}, \text{ or } v_2 \text{ c.c., suppose.}$$

Now find the density of the mixture, D, as in Experiment 29.

The mass of the mixture is $\mu - m_2$ grs. and its volume is

$$\frac{\mu - m_2}{D}, \text{ or } v \text{ c.c., suppose.}$$

The contraction therefore is $v_1 + v_2 - v$ c.c.

Now we must find the percentage of Absolute Alcohol in the mixture. If the percentage of Absolute Alcohol in the Alcohol used is found as in Experiment 38 to be π, the mass of Absolute Alcohol in the $m_1 - m_2$ grs. is $\dfrac{\pi(m_1 - m_2)}{100}$ grs., therefore the percentage in the mixture is $\dfrac{\pi(m_1 - m_2)}{\mu - m_2}$ grs. Thus we have found that for a mixture of Absolute Alcohol and water containing $\dfrac{\pi(m_1 - m_2)}{\mu - m_2}$ % by mass of the former the contraction at $t°$ is $v_1 + v_2 - v$ c.c.[42]

Enter your results in a tabular form as follows :—

	Mass in grs.	Density in grs. per c.c.	Volume in c.c.	
Absol. Alcohol	...	...	...	μ = ... grs.
Water . . .	...	...	...	m_1 = ... grs.
Mixture . . .	...	...	...	m_2 = ... grs.

μ = ... grs.
m_1 = ... grs.
m_2 = ... grs.
t = ... C
π = ... %

Contraction =   ...   % of Absolute Alcohol = ...

Repeat this experiment, using different percentages of Alcohol and Water. Finally, draw a curve showing the relation between the contraction and corresponding percentage of Absolute Alcohol in the different mixtures. It will be found that the maximum contraction takes place when the mixture contains about 46% of Absolute Alcohol.

Find as above the contractions which take place on mixing strong sulphuric acid and water in different proportions.

[42] We have neglected the contraction that took place due to the water originally present in the Alcohol we use.

H. Barometers.

A barometer is an instrument by which the pressure exerted
by the atmosphere may be measured.   A clean glass tube,
nearly a metre in length, is closed at one end and filled completely with pure dry mercury.   It is then inverted into a basin
of the same liquid.   The mercury in the tube drops a little
leaving a vacuous space above it, and the vertical height of the
column, when it comes to rest, above the level of the mercury
in the basin, is supported by and therefore indicates the pressure
of the atmosphere.   The pressure, due to the weight of this
column, on the sectional area of the tube on the same level as
the surface of the liquid in the basin is equal to the pressure of
the atmosphere on an equal area.   Suppose the sectional area
of the tube is 1 sq. cm. and the height of the column h cm.,
the pressure on this area is equal to the weight of h c.c. of
mercury, or hd_t grams-weight, where d_t is the density of mercury
at the temperature, or is equal to a force of $hd_t g$ dynes.
This therefore is the pressure of the atmosphere per square centimetre.   Since, in a liquid, pressure on a given area is trans-
mitted equally in all directions to all equal areas, the section of
the tube, as long as it is not very small, makes no difference[43]
to the height of the mercury column at a given time.   In whatever units we express the height of the column, we must use the
corresponding areal unit to express the area.   Thus, when we
incorrectly say the pressure of the atmosphere is equal to 75 cm.,
we mean that on every sq. cm. it exerts a pressure equal to the
weight of 75 c.c. of mercury.

*40· To make a barometer.

Apparatus.   A clean glass tube about ·75 in diameter and
a metre in length : Mercury Retort-Stand and Clamp : small
strong vessel : Evaporating Dish : Bunsen Burner : Dilute
Nitric Acid : strong Glass Flask : Funnel : separating Funnel.

Experiment.   In order to make a barometer which shall give

---

[43] Except the correction for capillarity which we neglect.

fairly accurate readings a number of precautions must be taken.

(1) The mercury must be pure, and clean and dry, for mercury, that is not so, not only has a different density but sticks to the glass. Place the mercury in a strong glass flask and pour some very dilute Nitric Acid upon it. Mix the two together for some considerable time until the mercury breaks up into tiny globules. Then pour the Nitric Acid off and replace it by some pure water. Clean the mercury with water in a similar way. The water must be several times renewed until it contains no trace of acid as tested by Litmus paper. Decant off the excess of water or separate the mercury by a separating funnel. Now squeeze the mercury through a clean piece of linen and finally filter it through a filter paper in which have been made a number of small holes by a needle-point. Place the mercury in an evaporating dish and heat it on a sand bath to get rid of the last traces of moisture.

(2) The glass tube, before the end is closed, must be thoroughly clean and dry. Wash it successively with boiling Nitric Acid, Distilled Water, Caustic Soda, Distilled Water, and dry it thoroughly. Then close the end in a blowpipe flame.

(3) When we fill the tube, both it and the mercury must be hot. To the shank of a clean funnel attach by a piece of india-rubber tubing a narrow glass tube reaching almost to the bottom of the barometer tube. Fill the tube by pouring mercury into the funnel. When the barometer tube is being filled occasionally shake or tap it to get rid of any air bubbles. When quite full, close the end by the finger and invert it in a basin containing the rest of the clean mercury. The space above the top of the mercury column ·in the tube will even now not be absolutely free from air. To get a Torricellian vacuum requires a long and tedious process of boiling the mercury. The barometer made as above will be sufficiently accurate for our purpose. The height of the barometer is the *vertical* height of the top of the column above the level of the liquid in the basin. If we read this vertical height with a separate metre rule, it is only a matter of convenience to arrange the tube itself

vertical by a plumb line, but if the scale is fixed to the tube it is evidently necessary that the tube itself should be accurately vertical. The space above the column in the tube is never really a perfect vacuum, even if all the air is got rid of. It is filled with mercury vapour, the pressure of which is however so small at the ordinary temperatures that its influence on the true height may be neglected.

A barometer made as above is not portable and is very liable to be broken and the mercury to be spilled. The syphon barometer is a portable and, reliable instrument and can be readily fixed to its position on the wall of the room, and is inexpensive withal. Good and simple directions for making one are given in Guthrie's *Practical Physics*, p. 132.

41. The Portable Barometer.

Apparatus. Glass tube about 40 cms. long and ·3 cm. in diam.: Pure Mercury: Half-metre Rule.

Experiment. We can find the pressure of the atmosphere in a very simple way as follows. Clean and dry the glass tube thoroughly and with the blowpipe flame close one end of it. Place the other end under mercury and heat the tube, driving out some of the enclosed air. Take away the burner and let the tube cool, still keeping its end under the mercury. A column of mercury rises in the tube. When it has cooled to the temperature of the room, lift the tube up vertically, being careful not to handle it more than is necessary, otherwise the heat of the hand will cause the air to expand. The column of mercury will be supported by the excess of the pressure of the atmosphere over that of the enclosed air. Now measure the length, l, of the air column and the length, h, of the mercury column. The enclosed air is now at a pressure less than the atmosphere by the weight of the mercury column. Now turn the tube upside down and again measure the length, l', of the air column. The air now is at a greater pressure than the atmosphere by the weight of the mercury column. Since the temperature between our measurements is supposed not to have altered, we have by Boyle's Law [Experiment 42], if H is the required atmospheric pressure,

$$l(H-h) = l'(H+h);$$

$$\therefore \quad H = \frac{h(l+l')}{l-l'}.$$

Repeat these measures twice more and take the mean of your results as the correct value of the atmospheric pressure. Compare it directly with the barometer reading.

42. To prove Boyle's Law (Fig. 12).

Apparatus. Two glass tubes of uniform bore, one about 20 cm., the other about 60 cm. long, and ·75 cm. in diameter: a length of india-rubber tubing about 30 or 40 cm. long, of somewhat less diameter [44]: clean Mercury: Metre Rule: Curve Paper: a board about 10 cm. × 60 cm., to which the apparatus is to be fixed: a piece of Cardboard about 30 cm. × 2 cm.: Mercury.

Experiment. Clean and dry the glass tubes. Close the end of the shorter one, *A*, in the blowpipe flame, and proceed to graduate it as follows [45]:—Fix it with its closed end downwards vertically along the middle of the piece of cardboard. Mark where the closed end is, and place o against the mark. Now, pouring into the tube equal volumes of mercury at a time, mark the levels at which the mercury stands. Take off the tube from the cardboard, empty the mercury out, and fix to its open end one end of the india-rubber tubing, binding it on with wire to make the attachment firm. Warm the mercury as well as the glass tube to get them quite dry, and then fill the glass and india-rubber

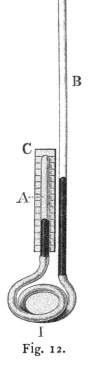

Fig. 12.

---

[44] Thick india-rubber tubing lined with canvas ought to be used to withstand the pressure. We may replace the india-rubber tubing by a *T* tube fitted to an india-rubber ball as explained in *Practical Work in Heat*, Experiment 27.

[45] It may be graduated as explained in Experiment 19, using mercury in place of water.

tubes full of the warm mercury. To the free end of the india-
rubber tube fix tightly on, as above, the longer glass tube, *B*.
Now turn the closed end of the shorter tube upwards and, closing the longer tube with the finger, lower it to allow sufficient
air to pass into the former so as to occupy rather more than
half its volume. Nail the cardboard with its longer side vertical
on the wooden support and firmly fix the shorter glass tube
parallel to it with its closed end coincident with the zero mark
on the cardboard. Adjust two wire guides by the side so that
the longer tube may be moved up and down in a parallel
direction. The upper part of this tube ought to pass through
a cork, so that, when not held in the hand, the wire guide may
prevent it from falling. We have now a volume of air in the
closed tube, which we can read off by the cardboard scale, and,
by raising the longer tube up or down, we can increase or
decrease the pressure of the enclosed air. This apparatus
should be fixed in a permanent position on the wall of the
Laboratory.

Boyle's Law states that if we alter the volume of a given mass
of air by altering its pressure, the volume will decrease or
increase in exactly the same proportion that the pressure
increases or decreases, provided the temperature of the air is
kept constant. In other words, the pressure, *P*, of a gas varies
inversely as its volume, *V*, if the temperature remains constant, or
in symbols

$$P \propto \frac{1}{V};$$

$$\therefore \quad P = \frac{k}{V}, \text{ where } k \text{ is a constant,}$$

$$\text{or} \quad PV = k,$$

that is if we, under these conditions, in any way alter the pressure
and the volume, the product of the two will be found to be
constant.

The pressure of the enclosed air is evidently equal to the
atmospheric pressure increased by or lessened by the difference
of the heights of the mercury columns in the two tubes, according as the height of the column in the closed tube is below or

above that in the other tube[46]. Lower the open tube as far as it will go and note carefully the volume, V, of the enclosed air. Measure the difference, h, of the heights of the mercury column and read the barometric height H. Move the open tube upwards for about three centimetres. Again take the readings of the volume and the heights. Continue moving the tube and taking readings until you can raise the tube no more, and enter your results in a tabular form as follows :—

Barometric height $H = \ldots$ cm. $t = \ldots$.

Volume of enclosed air V in c.c.	Difference in height of the two mercury columns h in cm.	Pressure of the enclosed air, $H \pm h$.	Pressure × Volume.
... ...	... ...	... ...	...

The numbers in the last column ought to be constant.

Now plot your results on a curve[47], taking the pressures as abscissae and the volumes as ordinates. This curve is called the isothermal curve for air at this temperature and will be found to be a rectangular hyperbola. By surrounding the shorter tube with a water jacket we can find the isothermals for any given temperature by heating the water to this temperature which must be kept constant during our experiment.

43. To measure the pressure of the water and the gas in the Laboratory.

Apparatus. A U-tube as used in Experiment 31 : India-rubber tubing to connect it with the mains : Half-metre Rule : Mercury.

i. *To measure the pressure of the water.*

Attach one arm of the U-tube, containing mercury to a height of

---

[46] When the pressure is altered by raising or lowering the tube, time should be allowed for the enclosed air to regain its original temperature before taking readings.

[47] See Appendix B 2.

about 20 cm., to the water-tap by means of the india-rubber tubing.   Turn the tap on slowly and measure the difference, h, of the levels of the mercury in the two arms.   This is the height of a mercury column whose weight is equivalent to the pressure of the water supply.   If d_t is the density of mercury at the temperature of the room, the height of the water-level above the top of the mercury in the shorter column is equal to hd_t cm.; and evidently its pressure is hd_t grams-weight per sq. cm.

ii. *To measure the pressure of the gas.*

Replace the mercury by water in the U-tube, and attach it to a gas nozzle and proceed as before, expressing the pressure in grams. weight per sq. cm.

I. The Simple Pendulum.

44. To find the relation between the length of a pendulum and its time of vibration.

Apparatus. A heavy metal bob about 2 cm. in diameter, attached to a 'fine wire about 120 cm. in length : Metre Rule : Watch : Pair of Pliers : Retort Stand : Clamp.

Experiment. Suspend the bob by the wire from a fixed support, or place the end between the jaws of a pair of pliers, fixed vertically by a clamp screwed tightly to a retort stand, which also ought to be fixed firmly to the table.   Tie a thread round the middle of the bob and draw it a very small angle out of the vertical.   Start the pendulum swinging by burning the thread.   This method prevents the bob at the same time rotating. The time of a complete vibration is the time taken by the pendulum to make two successive passages past a given point *in the same direction*, i. e. to make a swing-swang.   Now by the watch determine as accurately as possible the time it takes for the pendulum to make twenty complete vibrations, starting to count when it is in its lowest position, that is, when it is moving with the greatest velocity. Repeat this observation twice more, and take the average of results. Dividing this number by twenty, we get the time, t, of one complete vibration in seconds. Measure carefully the length of the wire from its support to the top of the spherical bob, and to this length add the radius of the bob measured as in Experiment 3, which gives us the length, l, of the pendulum, and repeat the observations twice more, each time shortening the wire, and enter your results in a tabular form as follows :—

t in seconds.	t^2.	l in cm.	$\dfrac{l}{t^2}$.	g in cm. per sec. per sec.
...	...	...	...	
...	...	...	...	...
...	...	...	...	...

Mean =   ...

The numbers in the fourth column, obtained by dividing the length of the pendulum by the square of the corresponding time of the vibration, will be found to be a constant number, showing that the square of the time of vibration is proportional to the length of the pendulum.

*45· **To determine g, the acceleration due to the force of gravity, and the length of the pendulum that locally beats seconds.**

Apparatus. As in Experiment 44.

Experiment. The formula connecting t and l, when the angle through which the pendulum swings is very small, is

$$t = 2\pi \sqrt{\frac{l}{g}},$$

where g is the acceleration due to the force of gravity, t the time of a complete vibration of the pendulum, and l its length ;

$$\therefore \quad g = \frac{4\pi^2 l}{t^2}.$$

F

Taking $\pi = 3.1416$, calculate for each series of observation in the last Experiment a value of g, and enter them in the last column. Find from Appendix A. 6 the true local value of g, and determine the percentage errors of your results.

The above equation is only true for a really *simple* pendulum, i. e. a pendulum in which the mass of the bob is concentrated into a point. In order to correct for the error due to our large spherical bob we ought to have increased l by $\dfrac{2r^2}{5l}$, where r is its radius. This correction is larger, the shorter the length of the pendulum, but the error due to neglecting this correction in this Experiment is less than the probable error of the result obtained, and therefore we need not take account of it. We have neglected also the resistance of the air for a similar reason.

It is evident that the length of the pendulum that beats seconds, i. e. which performs a complete oscillation in two seconds, may be obtained by putting $t = 2$ in the above equation, and therefore is equal to $\dfrac{g}{\pi^2}$ cm. Look in the tables for as near a correct value for g as possible for the latitude you are in, and calculate the length of the seconds pendulum. Arrange your wire of the calculated length and compare the rate of vibration with the time indicated by a watch by noting the time of fifteen complete vibrations. If they do not take exactly thirty seconds, either shorten or lengthen the wire accordingly until you have the correct length.

J. Capillarity, etc.

That the superficial film of a liquid is in a different physical state to the rest of the liquid follows from the fact that the molecules in the interior of the liquid are attracted by their surrounding neighbours equally in every direction, whereas the molecules which are at a distance from the surface less than the diameter of the sphere of molecular attraction have an unbalanced attraction towards the interior. The surface of a liquid acts, therefore, as if it were in a state of tension, i. e. each part of it tends to contract. If we imagine any line drawn in the surface, each unit of length is in equilibrium under two equal and opposite forces due to this tendency of the surface on each side to contract. The force thus exerted across unit length in the surface is called the *surface tension* of the liquid, and is measured in dynes per linear centimetre. Thus a soap film acts like a stretched sheet of india-rubber—the only difference being that the tension of india-rubber varies with the amount of stretching, whereas the tension of a liquid film is the same however the film may be extended, provided the temperature remains constant.

The force of adhesion to glass of a liquid, such as water, which wets it, is greater than the force of cohesion of the molecules of the liquid among themselves, and therefore the surface of water in a glass-vessel is concave—the narrower the vessel the more easily is this seen to be so. In consequence of the surface tension the pressure at the concave surface is less than at the horizontal surface of the liquid. Therefore on plunging a glass tube in water, the water will be seen to rise in it to a certain height above the outside level so as to preserve hydrostatic equilibrium—the narrower the tube, the more pronounced is the concavity, hence the greater the diminution of pressure, and therefore the higher will the water rise. The direction of the surface tension at each point of the contact line of the liquid and the glass is tangential to the surface and perpendicular to the line of contact. The angle at which a liquid meets a solid surface has been proved to be constant for the same liquid and solid under the same conditions, and is called the 'angle of capillarity.' In the case of a liquid such as mercury, which does not wet glass, and in which the force of cohesion among the liquid particles is greater than the force of adhesion to the glass, the surface is convex, and as a consequence the surface tension acts so as to depress the mercury in a glass tube below the outside level.

In Fig. 13, which illustrates the rise of a liquid in a narrow glass tube, θ is the angle of capillarity, and the tension T acts along the tangent to the concave surface.

* **46. To determine the value of the surface tension of soap solution.**

Apparatus. Copper wire about No. 18 B. W. G.: Soap Solution: Evaporating Dish: Balance and Ledge.

Experiment. Cut off and clean about 9 cm. of the copper wire: bend it into three sides of a rectangle, the two parallel legs being about 3 cm. long. Measure carefully the length, *l* cm., of the top cross piece. Hang the wire fork by a hair from the hook at the end of the left arm of the balance, so that the cross piece is horizontal. Now arrange an evaporating dish containing the soap solution on the ledge so that the two legs of the wire fork dip into the soap solution to a depth about one cm. Raise the dish for a moment so that the fork may be entirely wetted by the solution and break any film that may be formed. Counterpoise with sand the partly immersed fork. When this has been accurately done, tip the balance beam so that the whole fork may be immersed in the soap solution. The counterpoise will now be found unable to restore equilibrium. Carefully add masses until the balance returns to its original position of equilibrium. A soap film will be drawn up, and, if we neglect the weight of the film, the weight of the mass, *m*, we have added, i. e. *mg* dynes, is equal to the stretching force of the film. The film has two surfaces, and if T is the surface tension in dynes per linear centimetre at the temperature, *t*, of the solution, its stretching force is equal to $2\,Tl$, where *l* is the breadth of the film above measured

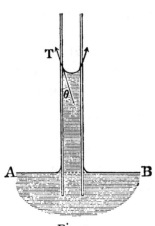

Fig. 13.

$$\therefore\quad 2\,Tl = mg,$$

$$\text{or}\quad T = \frac{mg}{2\,l}\ \text{dynes per lin. cm.}$$

Repeat the above experiment with pure distilled water [48] and glycerine, being very careful to clean the wire fork thoroughly before changing the liquid.

47. To find the relation between the radius of a capillary tube and the height to which a liquid rises in it, and hence to determine the surface tension of the liquid.

Apparatus. The three capillary tubes, whose radii were measured in Experiment 11 : Scale etched on glass as in Experiment 1 : Beaker : Thermometer : Clamp.

Experiment. Wash and dry the tubes thoroughly. Bevel off one end of the glass scale to a point, and along the scale parallel to its longer side fix by wax one of the capillary tubes, so that one end of it projects about a centimetre or so beyond the point. Now, by means of a plumb-line, support the tube quite vertically in a clamp, arranging it so that the pointed end of the glass scale just touches the surface of distilled water in the beaker. Draw up by suction the water in the tube two or three times and then observe carefully the height at which the water comes to rest in the tube above the outside level. Reverse the ends of the capillary tube on the scale, repeat the above experiment, and take the mean of the two readings as the correct height, *h*. Do exactly the same with the other two capillary tubes. Note the temperature, *t*, of the water and enter your results in a tabular form as follows :—

$$t = \ldots \,°C.$$

No. of tube.	Radius in cm. (r)	Mean height in cm. (h)	$h \times r$.	Surface-tension in dynes per lin. cm. (T)
1	...	...	...	
2	...	...	...	...
3	...	...	...	...

[48] In order to get a fair-sized film of water, the water must be quite free from contamination, as the slightest grease lowers the surface tension considerably. The water should be drawn from a wash-bottle recently filled with pure distilled water.

If the experiments have been performed carefully, it will be found that the product of the height (h) by the radius (r) of the corresponding capillary tube is a constant number ; in other words, the height to which a liquid rises in a capillary tube varies inversely as the radius of the tube.

In order to determine the surface tension of water by this experiment, we note that it acts tangentially to the concave surface of the water all round the circle of contact with the glass. If T is the surface tension per cm, $2\pi r T$ is the tension around the circle of contact, where r is the radius of the tube. The vertical component of this force, tending to raise the water, is $2\pi r T \cos\theta$, where θ is the angle of capillarity. This is in equilibrium with the weight of the raised column of water. If h is the height to which the water has risen above the outside level, the mass of the column is $\pi r^2 h \Delta_t$, where Δ_t is the density of water at the observed temperature $t°$, and therefore its weight is $\pi r^2 h \Delta_t g$ dynes, where g is the acceleration due to gravity.

$$\therefore \quad 2\pi r T \cos\theta = \pi r^2 h \Delta_t g ;$$

$$\therefore \quad T = \frac{r h \Delta_t g}{2 \cos\theta}.$$

In the case of liquids that wet glass θ is so small as to be negligible, $\therefore \cos\theta = 1$,

$$\therefore \quad T = \frac{r h \Delta_t g}{2} \text{ dynes per linear centimetre.}$$

Calculate from the results of your experiment the surface tension of water, and enter the values in the fifth column of the above table.

Repeat the above experiments with alcohol, ether, and turpentine, and calculate their surface tensions.

48. To compare the cohesion of different liquids by their drop-mass.

Apparatus. A glass bulb about 4 cm. in diameter with a short stem : a glass funnel with a stop-cock : Retort Stand and Clamp : Beaker : Balance : Watch.

Experiment. Carefully clean the glass bulb and support it, after dipping it in water, in the clamp. Arrange the funnel so that its stem is about 1 cm. above the glass sphere. Counterpoise the beaker. Fill the funnel with distilled water and open the stopcock so that the water will flow out at such a rate upon the sphere as to drop from its under surface at the rate of two drops per second. Now catch a hundred of these drops in the beaker and determine their mass. Make two more determinations and take the mean of the results as the mass, m, of a hundred drops. Now alter the drop rate to one in 5 seconds, and then to one in 10 seconds, and in each case determine as above the mass of a hundred drops. Repeat the above experiments, using successively Alcohol, Glycerine, Benzol, Turpentine, and enter your results in a tabular form as follows:—

Mass of 100 *drops in grs.*

Radius of sphere $= \dots$ cm.

Name of Liquid.	Drop-rate.			Density of Liquid.
	2 in 1″.	1 in 5″.	1 in 10″.	
Water . .	...	...	...	...
Glycerine .	...	...	...	...
Benzol . .	...	...	...	...
Turpentine	...	...	...	...
Alcohol .	...	...	...	...

The effects of density and cohesion are opposed to each other, but the results of experiment show that an increase in density, which of itself tends to diminish the size of the drop, may be more than counterbalanced by increased cohesion.

How is the drop-mass affected by altering the drop-rate? Ascertain the effect on the drop-mass by using spheres of different radii.

When a gas flows through a small aperture in a thin metal plate the process is called *effusion*, and the velocity of efflux varies as the square root of the height, H, of a homogeneous atmosphere of the gas at o°C which would exert the normal barometric pressure due to 76 cm. of mercury. This height evidently varies inversely as the density, d, of the gas, and therefore the velocity of efflux, v, varies inversely as the square root of the density.

$$\therefore \ v \propto \frac{1}{\sqrt{d}}.$$

In any gas this velocity is not affected by changes of pressure, for H varies directly as the pressure p and inversely as the density d, $\therefore H \propto \dfrac{p}{d}$, but by Boyle's Law $\dfrac{p}{d}$ is constant at the same temperature. Thus H is constant, and therefore v the velocity of effusion of each gas is constant under all circumstances of pressure.

* 49. To compare the densities of gases by their rates of effusion.

Apparatus. A glass tube 2 cm. in diameter and over 30 cm. in length : Piece of thin Platinum Foil about 2·5 cm. square : Retort Stand and Clamp : Mercury : a tall glass Jar [49] about 5 cm. in diam. and 16 cm. high : Watch.

Experiment. Grind one end of the glass tube with emery powder and turpentine on a plane surface until it is perfectly flat. With a very fine needle pierce as small a hole as possible in the centre of the piece of platinum foil. The hole may be made smaller by a gentle tap with a hammer, the foil lying on a smooth surface. Fix the foil by means of beeswax to the ground edge of the glass tube. Now cut off a small piece of glass rod and draw it out with a fine stem about 15 cm. long. Along the stem fix a little melted beeswax at distances about 3 cm. apart. Fill the glass jar about 12 cm. high with mercury. Now take the glass tube and fill it with hydrogen gas over mercury, closing the small hole in the platinum foil, if necessary with the finger. Introduce the glass stem and invert the tube filled with hydrogen under the mercury in the jar and clamp it

---

[49] A Hydrometer jar in which the upper portion is of larger diameter than the lower, economises the use of the mercury.

vertically so that it is half immersed. By the watch note the times at which the 'marks on the stem reach the surface of the mercury as the gas is pressed through the small hole. Now fill the tube with oxygen and repeat a similar series of observations. The densities of the two gases vary as the squares of the velocities of effusion, i.e. inversely as the squares of the times which a given division on the glass stem takes to pass above the surface of the mercury. Enter your results in a tabular form as follows, where T_1, T_2 ... are the times taken by the successive lengths on the glass stem to pass the surface of the mercury in the case of Hydrogen, and T_1', T_2' ... the corresponding times for Oxygen.

Hydrogen.	Oxygen.	Density of Oxygen referred to Hydrogen.
T_1	T_1'	$\dfrac{T_1'^2}{T_1^2}$
T_2	T_2'	$\dfrac{T_2'^2}{T_2^2}$
...	...	...
...	...	...

Mean =   ...

Compare as above the densities of Carbon Dioxide Gas, and Nitrogen with that of Hydrogen.

*50· To determine the solubility of a solid in water.

Apparatus. A flask of about 300 c.c.: a smaller flask of about 80 c.c. : Balance : Bunsen Burner : Thermometer : 'Crystallised Sodium Sulphate.

Experiment. Half fill the larger flask with distilled water and place the thermometer in it. Boil the water and add, in a powdered state, an excess of the solid until there remains some undissolved. Take the flask away from the burner and let it cool. Counterpoise the smaller flask with a tare mass μ of about 150 grams ; when the solution has cooled to about

80° drop into it a small crystal of Sodium Sulphate so as to precipitate any of the salt held in solution in excess of the normal quantity at the saturation point. Read the temperature, t, of the solution and pour some of it into the small flask, and taking away the tare mass, add known masses, m_1, to restore equilibrium. Then $\mu - m_1$ is the mass of the solution. Place the small flask on a sand bath and evaporate off all the water, being careful not to hasten the evaporation for fear of losing any of the salt by 'spitting.' When thoroughly dry replace the flask on the scale pan and restore equilibrium by known masses m_2. Then $\mu - m_2$ is the mass of the salt that saturated the solution at $t°$, and $m_2 - m_1$ is the mass of the water present.

Calculate now by the formula $\dfrac{\mu - m_2}{m_2 - m_1} \times 100$ the mass of salt required to saturate 100 grams of water at this temperature, i.e. its solubility, s. Repeat this experiment when the original solution has cooled to 60°, 40°, 35°, 30°, 20° successively, and arrange your results in a tabular form as follows :—

$$\mu = \ldots \text{grs.}$$

$t°$.	$\mu - m_1$.	$\mu - m_2$.	s.
...	...	...	...
...	...	...	...
...	...	...	...

Now plot a curve showing the relation between the mass of salt required to saturate 100 grs. of water (i. e. s) and the corresponding temperature (t). See Appendix, B. 2. It will be found that at the temperature 33° the solubility of Sodium Sulphate is a maximum.

Find as above the solubility of the following salts in water, drawing a curve for each on the same piece of curve paper: Nitre, Sodium Chloride, Potassium Chloride, Copper Sulphate.

APPENDIX A.

$t°$ C.	(1) Volume of unit mass and density of water at different temperatures.		(2) Volume of 1 c.c. at 0° and density of mercury at different temperatures.	
	Volume of unit mass.	Density.	Volume of 1 c.c. at 0°.	Density.
0	1·0001	0·99988	1·0000	13·596
4	1·0000	1·00000	1·0007	13·586
10	1·0003	·99976	1·0018	13·572
15	1·0008	·99917	1·0027	13·559
20	1·0017	·99827	1·0036	13·547
25	1·0029	·99713	1·0045	13·535
30	1·0043	·99578	1·0054	13·523
35	1·0059	·99407	1·0063	13·511
40	1·0077	·99236	1·0072	13·499
45	1·0097	·99029	1·0081	13·487
50	1·0120	·98821	1·0090	13·474
55	1·0144	·98580	1·0099	13·462
60	1·0169	·98339	1·0108	13·450
65	1·0197	·98067	1·0117	13·438
70	1·0226	·97795	1·0127	13·426
75	1·0257	·97495	1·0136	13·414
80	1·0289	·97195	1·0145	13·401
85	1·0322	·96876	1·0154	13·389
90	1·0357	·96557	1·0163	13·377
95	1·0394	·96212	1·0172	13·365
100	1·0432	·95866	1·0182	13·353

(3) *Densities of Solids.*

(a) *Elements:*—

Bismuth . . . . 9·60 to 9·88
Copper (sheet) . . 8·85 to 8·89
  „ (wire) . . 8·30 to 8·89
Graphite . . . . 2·10 to 2·58
Iodine . . . . . 4·95

(β) *Compounds and Mixtures:*—

Alum. . . . . . 1·72
Amber . . . . . 1·08
Beech . . . . . . ·70
Beeswax . . . . . ·96
Boxwood . . . . ·94

Elements (continued).

Iron (wrought) . .	7·60 to	7·79
„ (cast) . . .	7·50	
„ (wire) . . .	7·60 to	7·73
„ (steel) . . .	7·80 to	7·90
Lead	11·07 to	11·40
Phosphorus . . .	1·83	
Platinum (foil) . .	21·16 to	21·31
„ (wire) . .	21·16 to	21·53
Potassium	·875	
Silver	10·43 to	10·53
Sodium	·974	
Sulphur (roll) . .	1·98 to	2·02
Tin	7·23 to	7·37
Zinc	6·86 to	7·20

Compounds, &c. (continued).

Brass	7·8 to	8·4
Camphor	·99	
Copper Sulphate (cryst.)	2·19	
Cork	·24	
Ebony	1·19	
Fluor Spar	3·20	
Galena	7·65	
Glass (crown) . . .	2·50 to	2·70
„ (flint) . . .	3·00 to	3·50
Granite	2·65	
Guttapercha . . .	·97	
India-rubber . . .	·93	
Iron Pyrites . . .	4·85	
Ivory	1·86	
Mahogany	·56 to ·85	
Marble	2·77	
Oak	·7 to 1·0	
Paraffin, M.P. 40° .	·872	
„ 45° .	·887	
„ 50° .	·908	
„ 55° .	·912	
„ 60°–80°	·923 to ·943	
Porcelain	2·40	
Quartz	2·63	
Salt (Rock)	2·01 to 2·20	
Sand (Dry)	1·42	
Slate	2·88	
Sodium Sulphate (cryst.)	1·46	
Tallow	·94	

(4) *Densities of Liquids at* 15° C.

Acetic Acid	1·05
Alcohol (Amyl)	·81
Benzine	·883
Benzol	·88
Chloroform	1·50
Copper Sulphate (15 % solution)	1·16
Ether	·72
Glycerine	1·26
Hydrochloric Acid	1·27
Mercury	13·559
Milk	1·03
Nitric Acid	1·55
Oil (Linseed)	·94
Oil (Olive)	·915

Rock Oil	. . . .	·85
Sodium Chloride (15 % solution)	. . . .	1·11
Sulphuric Acid	. . . .	1·84
Turpentine	. . . .	·87

(5) *Densities of Alcohol solutions at different temperatures.*

From Mendelejeff's Experiments, Pogg. Ann. Bd. cxxxviii (1869).

Mass % Absolute Alcohol.	Mass % Distilled Water.	Density at		
		10°.	15°.	20°.
100	0	·79789	·79368	·78946
95	5	·81292	·80863	·80434
90	10	·82666	·82247	·81801
85	15	·83968	·83544	·83116
80	20	·85216	·84793	·84367
75	25	·86428	·86007	·85581
70	30	·87614	·87200	·86782
65	35	·88791	·88378	·87962
60	40	·89945	·89537	·89130
55	45	·91075	·90679	·90276
50	50	·92183	·91797	·91401

(6) *Local values of g and of the 'seconds pendulum' at sea-level.*

The value of g in C. G. S. units at any place on the Earth's surface at sea-level is given by the formula

$$g = 980\cdot6056 - 2\cdot5028 \cos 2\lambda,$$

where λ is the latitude of the place.

	Value of g in cm. per sec. per sec.	Length of seconds pendulum in cm.
Equator .	978·1	99·103
Paris . .	980·94	99·390
Greenwich .	981·17	99·413
Berlin . .	981·25	99·422
Dublin . .	981·32	99·429
Manchester	981·34	99·430
Edinburgh .	981·54	99 451
Aberdeen .	981·64	99·461
New York .	980·19	99·314
Birmingham	981·25	99·422

(7) *Conversion Tables.*

1 centimetre	= ·394 inch.	1 inch	= 2·54 centimetres.
1 sq. centimetre	= ·155 sq. inch.	1 sq. inch	= 6·45 sq. centimetres.
1 cub. centimetre	= ·061 cub. inch.	1 cub. inch	= 16.386 cub. centimetres.
1 gram	= 15·432 grams.	1 grain	= ·065 grams.
1 kilogram	= 2·2 lbs.	1 lb.	= 453·593 grams.

If $t°$, $\theta°$ represent the same temperature on the Centigrade and Fahrenheit thermometers respectively,

$$\theta = \tfrac{9}{5}t + 32; \qquad t = \tfrac{5}{9}(\theta - 32),$$

at 39° F or 4° C a cubic foot of water weighs nearly as much as 62·4 lbs.

APPENDIX B.

1. Correction for the upthrust of air on the body weighed.

To correct for the upthrust of the air on the body weighed the following method will be found applicable to all cases, and is easily applied.

Let M be the apparent mass of the body in air or, in the case of Experiment 30, the upthrust of the given liquid on the glass stopper,

m the apparent mass of a volume of water equal to the volume of the solid;

(a) For solids m is either

(i) the loss of weight of the body in water;

or　(ii) the overflow of water from the density bottle ;

(β) For liquids m is either

(i) the mass of the water filling the density bottle ;

or　(ii) the loss of weight of the glass stopper in water ;

l the mass of air displaced by the body,

Δ_t the density of water at the temperature of the experiment.

1. Since when a body is weighed in air, by Archimedes' principle, it apparently loses weight equal to the weight of the air displaced, the true mass of the body is $M + l$ grs.

2. In a (ii) and β (i) it is easily seen for the same reason as above that the real mass of the volume of water is $m + l$. In

a (i) and β (ii) it must also be $m + l$, for the true mass of the body is l greater than its apparent mass in air.

3. Therefore the volume of the body is $\dfrac{m+l}{\Delta_t}$.

The density corrected for the upthrust becomes

$$d = \frac{M+}{m+l}\, \Delta_t \ \ldots\ldots \ (\text{i}).$$

If a is the density of air at the temperature and pressure of the experiment, the mass of air displaced by the body is

$$\frac{m+l}{\Delta_t}\, a = l;$$

$$\therefore \ \ l = \frac{ma}{\Delta_t - a} \ \ldots\ldots \ (\text{ii}).$$

Substituting this value for l in (i) we get

$$d = \frac{M}{m}\, \Delta_t - \frac{M-m}{m}\, a.$$

Almost always it will be sufficient to take ·0012 grs. for the value of a.

2. The graphical method of treating results of experiment.

If two quantities vary together and we require to know the relation of one to the other, we could not determine it for every value of the quantities. This would necessitate an infinite number of experiments. We therefore plot our results in a curve as follows :—Curve paper, on which the curve is to be drawn, is paper ruled into small squares. The best is that divided into square millimetres on which, for convenience, areas of square centimetres are also marked out in darker lines or lines of a different colour. It may be obtained from mathematical instrument makers, as well as a cheaper paper, in which the squares are not so small, and which will be as convenient for our purpose. Cut out a piece of curve paper of a suitable size, and on it draw two straight lines at right angles to one another, one close to and parallel to the bottom edge (the line of *abscissae*), the other close to the left-hand edge (the line of *ordinates*), their point of intersection being called the *origin.* In order to explain how a· curve is to be drawn let us refer to Experiment 42. In this case you have two columns of numbers, one column giving the volumes, the other the corresponding pressures of the air enclosed in the glass tube, and we wish to show how the two vary together.

Suppose the difference of the greatest and least volume is 15 c.c. and the difference of the greatest and least pressures is 70 cm. Along the line of abscissae mark off equal lengths to represent 2 c.c. The number of divisions on the curve paper taken for these lengths must of course depend on the number of divisions at your disposal, and be such that the total range of volumes may be included. Representing the smallest volume observed by the origin, place numbers at the mark to represent the corresponding volumes. Again along the line of ordinates mark off equal lengths (not necessarily of the same value as before) to represent 10 cm. of pressure, and taking the origin as your lowest pressure observed, place numbers at the marks to represent the corresponding pressures.

Along the line of abscissae place small dots at the points representing the volumes and along the line of ordinates at the points representing the pressures obtained in your experiment. Now mark with a very small cross the points obtained by the intersection of *imaginary* vertical lines through the dots representing the volumes with *imaginary* horizontal lines through the dots representing the *corresponding* pressures. With a piece of thin cardboard or a flexible lath as a guide draw carefully a fine continuous line through the average run of the series of points so obtained. The curve will not pass through all the points unless your observations have been perfectly exact. Some points will be above, some below, some upon the final curve.

The curve obtained, as above, for Boyle's Law, will be found to be a rectangular hyperbola, a property of which curve is that the rectangle under the coordinates of any point on it is a constant quantity. This is the same thing as saying that at any stage of the experiment $PV =$ constant. If a curve is found to be a straight line, it shows that the two quantities in question vary directly together. If we plotted the curve connecting the values of P and $\dfrac{1}{V}$ we should find it a straight line, since evidently Boyle's Law may be expressed as $P \propto \dfrac{1}{V}$, or the pressure varies inversely as the volume at a constant temperature. This graphic method not only points out where mistakes have been made by the observer, but also enables one to get approximate values for any intermediate points which have not been determined directly by experiment.

When a curve is completed write along the top of the paper the experiment to which the curve refers, and along the line of abscissae and the line of ordinates the names of the measurements to which each refer, stating also how many units each scale division represents [50].

3. To draw a perpendicular to a straight line from a given point without it.

Let AB be the given straight line, O the given point. On the other side of AB to the point O take any point C and

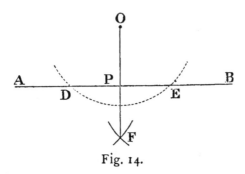

Fig. 14.

describe an arc of circle with centre O and radius OC cutting AB at D and E.

With centres E and D, and equal radii greater than half AE, describe two arcs cutting each other at F. Join OF cutting AB at P. Then OP is perpendicular to AB.

[50] For a figure of a curve see *Practical Work in Heat*, Fig. 1, by the author.

4. To draw a perpendicular to a straight line from a given point within it.

Let *AB* be the given straight line and *O* the given point. Measure off equal distances *OD* and *OE* on either side of

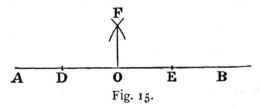

Fig. 15.

O and with centres *D*, *E* and equal radii greater than *DO* or *OE* describe arcs of circles cutting each other at *F*. Join *OF*. Then *OF* is perpendicular to *AB*.

APPENDIX C.

For those who have not the time or inclination to make the apparatus mentioned in this book the following price-list supplied by Mr. Groves, 89 Bolsover Street, Portland Place, London, W., may be of use. In many cases the apparatus is of two classes. That of Class B is made of varnished pine and has paper scales. That of Class A is of hard wood polished, has boxwood scales, and is of altogether superior workmanship.

	£	s.	d.
Boxwood Metre Scale divided into mm.		3	0
Boxwood Half-metre ,, ,, ,, ...		1	6
Steel rule 30 cm. long divided into mm.		2	0
Model Vernier in Boxwood		5	0
Sliding Callipers (Fig. 2)		6	6
Micrometer Screw-Gauge $\frac{1}{2}$ mm. pitch in box (Fig. 3) . .	1	2	0
,, ,, ,, 1 mm. pitch, Class B . . ,,		10	0
Strips of Glass per doz.			6
Beam Compass for Experiment 1		5	0
Opisometer (Fig. 5)		2	3
Two squared pieces of wood		1	6
Sphere of Ebonite		2	3
Cylinder of Ebonite		1	6
Circle pencil compass		1	0
Three Capillary tubes		1	6·
Metal cube and box		4	0
Density Bottle, drilled stopper (Fig. 9)		2	9
,, ,, mark on neck		1	0

		£	s.	d.
Hollow metal sphere			2	6
U-tube on stand, Class A (Fig. 10)			7	6
,, ,, ,, Class B			4	0
W-tube on stand, Class A (Fig. 11).			16	0
,, ,, ,, Class B			7	0
Apparatus mentioned Expt. 32. N.B.			7	0
Nicholson's Hydrometer			6	6
Tall glass jar for ditto			1	6
Hydrometer of variable immersion			3	6
Boyle's Law apparatus on iron stand, large tubes and boxwood scales, Class A		3	0	0
Boyle's Law apparatus (Fig. 12), Class B			13	0
Metal Pendulum bob			2	0
Three Glass bulbs			1	0

ADDENDA:

Page 25, line 12 from bottom, *instead of* about 1 cm. inside *read* the length of one side being about 2 cm.

As note on p. 27 add :—

[19] In this and the next two experiments find the volume also as follows : Half fill a glass vessel, such as a test-tube, of rather larger diameter than the body, with water, and mark its level with a piece of gummed paper. Carefully drop the body into the vessel. The volume of the water that rises above the mark is equal to the volume of the body. Weigh the vessel and its contents. Now with a pipette take away the water above the mark till it has returned to its original level. Weigh again. The difference of the masses employed, m suppose, is equal to the mass of the volume of water displaced. If Δ_t is the mass of 1 c.c. of water at the temperature of the experiment, $\dfrac{m}{\Delta_t}$ c.c. is the volume of the body.

Put the reference-note number 19 after

(1) the word 'value,' p. 27, line 15 from top ;
(2) the word 'sphere,' p. 28, line 3 from top ;
(3) the words '$\pi r^2 h$ c.c.,' p. 28, line 11 from bottom.

Page 31, line 6 from top : *between* glass *and* 60 *insert* tubing.

THE END.

CPSIA information can be obtained
at www.ICGtesting.com
Printed in the USA
BVHW040932010319
541456BV00034B/584/P

9 781330 872697